J.A. Pintozzi

The Apprentices Guide To Blueprint Reading

PURE VISION MACHINING

J.A. Pintozzi

COPYRIGHT

The Apprentices Guide To Blueprint Reading

1st ed.

TEXT COPYRIGHT© 2017 J.A. Pintozzi

ALL RIGHTS RESERVED

No part of this publication may be reproduced or transmitted in any form or by any means, electronic or mechanical including photocopying, recording or any information storage or retrieval system, without permission in writing from the author / Publisher J.A. Pintozzi japintozzi@gmail.com

The author of this book J.A. Pintozzi has been asserted in accordance with the Copyright, Designs and Patent act of 1988

Copyright © 2017 J.A. Pintozzi

All rights reserved.

The Apprentices Guide to Blueprint Reading

J.A. Pintozzi

DEDICATION
FOR ANTHONY

J.A. Pintozzi

CONTENTS

Introduction   i

PART 1
The Basics

1	The Components of a Blueprint	1
2	The Alphabet of lines	18
3	Multi-View Drawings	31
4	Sectional-View Drawings	40
5	Threads, Gears, Welding Symbols and Finish Marks	52

PART 2

GD&T

6	**Geometric Dimensioning and Tolerancing**	70
7	Tolerance	96
8	Features, feature Controls and Datum's	118
9	Form, Form Controls and Geometric Tolerancing	130
10	Machine Terms and manufacturing processes	161
11	Surface Finish, Geometric Deviations	174
12	Inspection Techniques	196
	Glossary	212
	Appendix	222

J.A. Pintozzi

INTRODUCTION

A Blueprint is a document that communicates a precise description of a part, including the intended size, shape, dimensions and any allowable tolerances. The description consists of pictures, words, numbers and symbols. A good blueprint will usually contain the geometry of the part, all critical functional relationships, tolerances, material, heat treat, surface coatings and part documentation such as part number and drawing revision number.
It did not take long for manufacturers to realize that drawing errors cost them time, material, customer satisfaction and MONEY. They realized the fact that, the farther the error traveled through the manufacturing processes, the greater their losses became. did not go unnoticed.

Blueprints transform ideas into products and communicate information among all parties involved, such as customers, sales, designers, drafters, purchasers, routers, machinists, machine operators, assemblers, and inspectors. A machinist must visualize the part from the blueprint before he or she can make it. After a part is made, an inspector usually checks the part to see if it conforms to the print. The inspector also needs to form a mental picture of the part from the print in order to inspect the part.

The interpretation of lines, symbols, dimensions, notes, and other information on a print is the other important part of print reading. Print reading is an essential skill for anyone working in the manufacturing industry. Accurate and satisfactory fabrication of a part described on a drawing depends upon the following:

• Correctly reading the drawing and closely observing all data on the drawing.

• Selecting the correct tools and instruments for laying out the job.

• Use the baseline or reference line method of locating the dimensional points during layout. thereby avoiding cumulative errors.

• Strictly observing tolerances and allowances.

• Accurate gagging and measuring of work throughout the fabricating process.

• Giving due consideration when measuring for expansion of the work piece by heat generated by the cutting operations. This is especially important when checking dimensions during operations, if work is being machined to close tolerances.

Why GD&T?

Before GD&T drawing errors compounded the cost of the error as the part moved further from initial design to production. Drawing errors cost:
- **Money**
- **Time**
- **Material**
- **Customer satisfaction**

To further complicate matters, there were disagreements over the drawings interpretation, as well as difficulty in communicating the drawings requirements. In some cases the machinists could not understand the designer's intent. This resulted in Bad parts passing Inspections and wasting money on parts that did not fit.

In this book we will be using both inch and metric dimensions. The drawings will state the units used and certain other information used in the drawing.

As you progress through this book you will notice that some practices are repeated. This was done because the concept is important and it may apply to more than one area of the book. It was also done to help reinforce the information in your mind.

CHAPTER 1
The Components of a Blueprint:

Blueprints (prints) are copies of mechanical or other types of technical drawings. The term blueprint reading means interpreting ideas expressed by others on drawings, whether the drawings are actually blueprints or not. Drawing or sketching is the universal language used by engineers, technicians, and skilled craftsmen. Drawings need to convey all the necessary information to the person who will machine, weld or assemble the object in the drawing. Blueprints show the construction details of parts, machines, ships, aircraft, buildings, bridges, roads, and so forth.

Original drawings are drawn, or traced, directly on translucent tracing paper or velum using a black ink, a pencil, or more commonly, a computer aided drafting system (CAD). The original drawing is kept as a master copy, and all necessary prints are produced from the master. These copies of the master are distributed to persons or offices where needed. Master prints that are properly handled and stored will last indefinitely.

The term *blueprint* is used loosely to describe copies of original drawings or tracings. One of the first processes developed to duplicate tracings produced white lines on a blue background; hence the term *blueprint.* Today, however, other methods produce prints of different colors. The colors may be brown, black, gray, or maroon. The differences are in the types of paper and developing processes used. A patented paper identified as BW paper produces prints with black lines on a white background. The diazo, or ammonia process, produces prints with either black, blue, or maroon lines on a white background.

- A good source of information regarding blueprints is the American National Standards Institute (ANSI) standards.

INFORMATION BLOCKS:
The draftsman uses information blocks to give the reader additional information about materials,
Specifications and so forth that are not shown in the blueprint or that may need additional explanation. The drafter may leave some blocks blank if the information in that block is not needed. The following paragraphs contain examples of information blocks.

TITLE BLOCKS:
The title block is located in the lower-right corner of all blueprints and drawings prepared according to ANSI standards. It contains the drawing number, name of the part or assembly that it represents, and all information required to identify the part or assembly. It also

includes the name and address of the organization preparing the drawing, the scale, drafting record, authentication, and date. A space within the title block with a diagonal or slant line drawn across it shows that the information is not required or is given elsewhere on the drawing.

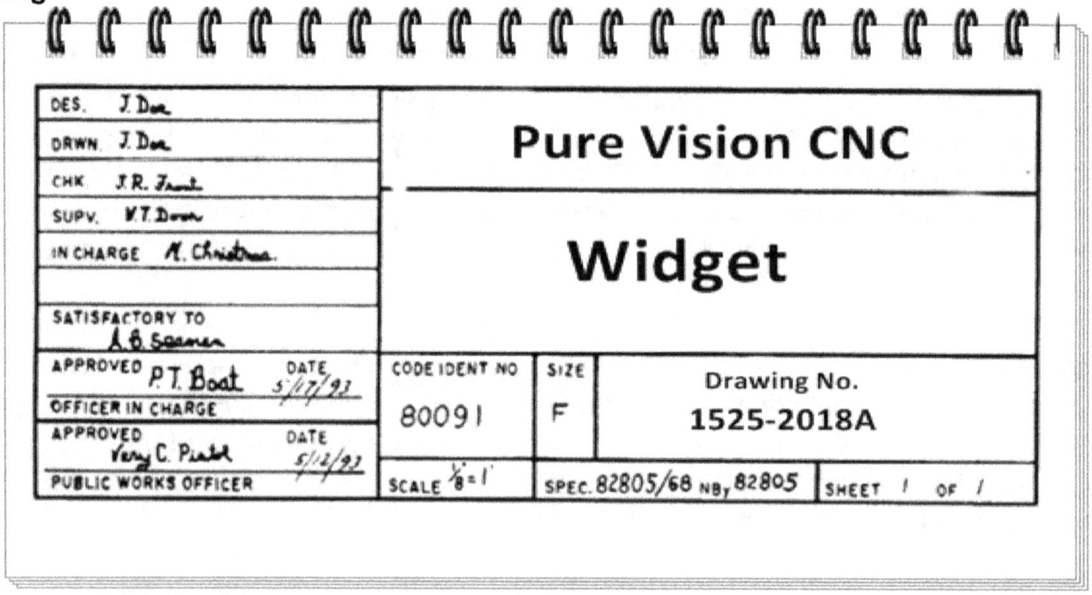

(Example of Title Block)

REVISION BLOCKS:

If a revision has been made, the revision block will be in the upper right corner of the blueprint, as shown in the figure below. All revisions in this block are identified by a letter and a brief description of the revision. A revised drawing is shown by the addition of a letter to the original number. When the print is revised, the letter A in the revision block is replaced by the letter B and so forth. As an example Drawing # 1525-2018B, would be a revised version of Drawing #1525-2018-A.

REVISIONS			
SYMBOL	DESCRIPTION	DATE	APPROVAL

THE BILL OF MATERIAL (BOM):

The bill of material block contains a list of the parts and/or material needed for the project. The block identifies parts and materials by stock number or other appropriate number, and lists the quantities requited. The bill of material often contains a list of standard parts, known as a parts list or schedule. BOMs may also contain additional information, such as materials, suppliers, and additional remarks. Depending on what standards are followed, a bill of materials may contain a "find number," a unique sequential identifier for the items. Such find numbers are referenced on the drawing using callouts to indicate which parts are which. For bills of materials that do not use find numbers, the assembly drawing indicates the item numbers directly.

ITEM Number	......DESCRIPTION...	FN	UNIT	Quantity

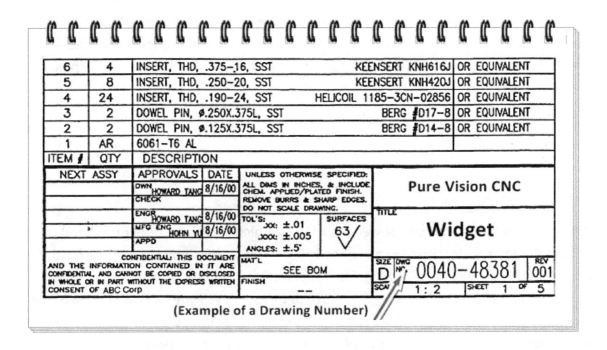

(Example of a Drawing Number)

DRAWING NUMBER:

Each blueprint has a drawing number, which appears in a block in the lower right corner of the title block. The drawing number can be shown in other places, for example, near the top borderline in the upper corner or on the reverse side at the other end so it will be visible when the drawing is rolled. On blueprints with more than one sheet, the information in the number block shows the sheet number and the number of sheets in the series.

REFERENCE NUMBER:

Reference numbers that appear in the title block refer to numbers of other blueprints. A dash and a number show that more than one detail is shown on a drawing. When two parts are shown in one detail drawing, the print will have the drawing number plus a dash and an individual number.

In addition to appearing in the title block, the dash and number may appear on the face of the drawings near the parts they identify. Some prints use a leader line to show the drawing and dash number of the part. Others use a circle 3/8 inch in diameter around the dash number, and carry a leader line to the part.

A dash and number identify changed or improved parts and right-hand and left-hand parts. On some prints you may see a notation above the title block such as "159674 LH shown; 159674-1 RH opposite." Both parts carry the same number. LH means left hand, and RH means right hand. Some companies use odd numbers for right-hand parts and even numbers for left-hand parts.

DRAWING NOTES:

Notes on a print are numbered, located near the title block, or in the upper left-hand corner of the print, they are identified with the heading "NOTES." They apply to either a portion of the print or the entire print, providing additional treatment, finish, processing, or other considerations. When it is necessary for a note to be associated with a specific area of the print, that note's identifying number is flagged and placed in the associated view, with a leader if appropriate. Several varieties of note flags are common in practice;

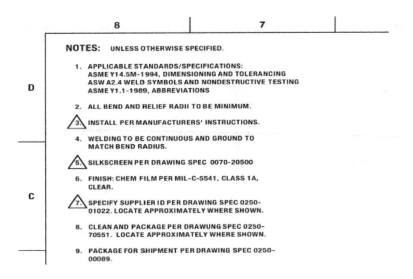

(Examples of different Note Flags)

ZONE NUMBERS:

Note the Zone Numbers around the Outside edge of the blueprint. The part in the image could be found by looking in zone B-2. Zone numbers serve the same purpose as the numbers and letters printed on borders of maps to help you locate a particular point or part. To find a point or part, you should mentally draw horizontal and vertical lines from these letters and numerals. These lines will intersect at the point or part you are looking for. You will use practically the same system to help you locate parts, sections, and views on large blueprinted objects.

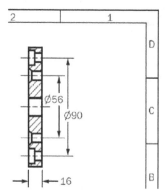

Parts numbered in the title block are found by looking up the numbers in squares along the lower border. Read zone numbers from right to left.

SCALE BLOCKS:

The scale block in the title block of the blueprint shows the size of the drawing compared with the actual size of the part. If the drawing is the same physical size as the actual part, the drawing is called "full size"; if it is not practical to make a drawing the same size as the part, the drawing is made smaller or larger.

The drawing is, therefore, scaled down or scaled up from the actual part. The scale may be shown as $1² = 2²$, $1² = 12²$, $1/2² = 1´$, and so forth. It also may be shown as full size, one-half size, one-fourth size, and so forth. The scale of a drawing may be any ratio, but the most common ones are 1:1, 1:2, 1:5, 1:10, etc. Scale may be shown in several different ways, such as FULL, 1 = 1, 1:1, 1/1, or some other similar form.

Most prints are made with all views drawn at the same scale; however, some prints may also contain views drawn at different scales. Scale is indicated on the title block and applies to all views in the drawing unless otherwise specified.

A line under a dimension indicates the dimension is not to scale. The line may be straight, or it may be wavy. As a rule, prints should never be scaled or measured to find a dimension; most are even marked with a note "DO NOT SCALE." Scaling prints may lead to serious problems for several reasons. The drawing may not be drawn to the correct scale, or the dimension may have been changed. This is less of a concern with CAD-generated drawings than with hand-drawn prints where a complete redraw due to a small design change was labor intensive and could be skipped. Regardless of how the print was generated, the paper may stretch or shrink.

SCALE BLOCK

If the scale is shown as $1" = 2"$, each line on the print is shown one-half its actual length. If a scale is shown as $3" = 1"$, each line on the print is three times its actual length. The scale is chosen to fit the object being drawn and space available on a sheet of drawing paper.

Never measure a drawing; use dimensions. The print may have been reduced in size from the original drawing. Or, you might not take the scale of the drawing into consideration. Paper stretches and shrinks as the humidity changes. Read the dimensions on the drawing; they always remain the same.

Graphical scales on maps and plot plans show the number of feet or miles represented by an inch. A fraction such as 1/500 means that one unit on the map is equal to 500 like units on the ground. A large-scale map has a scale of $1" = 10'$; a map with a scale of $1" = 1000'$ is a small-scale map. The following chapters of this manual have more information on the different types of scales used in technical drawings. The scale of a drawing is the ratio of the drawn object to its actual size. Scale is important because many times a sketch must be drawn smaller than the actual size of the object in order to fit the picture onto a manageable sheet of paper. A house blueprint, a wiring diagram, or a working drawing of a new car design are -typical examples of working drawings that usually are scaled down from-the actual size of the object. Conversely, drawings of small and complicated parts are drawn larger than actual size in order to emphasize required detail.

The scale used in a blueprint should always should be indicated on the print. The most frequently, used scales in the trades are the architect's scale, the mechanical engineer's scale and the clvil engineer's scale. While each tool looks very similar, the scale itself is graduated differently for each instrument. As seen Below:

Mechanical Engineers Scale	Architects Scale	Civil Engineers Scale
1/8 inch = 1" or 1 ft.	3/32" = 1'	1" = 1' or 10'
1/4 inch = 1" or 1'	3/16" = 1'	1" = 2' or 20'
3/8 inch = 1" or 1'	1/8" = 1'	1" = 3' or 30'
3/4 inch = 1" or 1'	1/4" = 1'	1" = 4' or 40'
1/2 inch = 1" or 1'	3/8" = 1'	1" = 5' or 50'
1 inch = 1 ft.	1/2" = 1'	1" = 6' or 60'
1 1/2" = 1" or 1'	3/4" = 1'	
2" = 1" or 1'	1" = 1'	
3" = 1" or 1'	1 1/2" = 1'	
4" = 1" or 1'	3" = 1'	

In the manufacturing industry, the machinists rule, is used most frequently. It is especially well suited for working with dimensions in fractions of an inch.

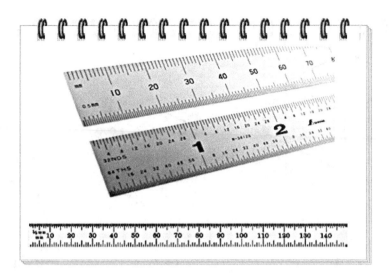

SCALE:

The scale of a drawing is the relationship between the size of the drawing and the size of the actual part. If the drawing is the same physical size as the actual part, the drawing is called "full size"; see the image below. If it is not practical to make a drawing the same size as the part, the drawing is made smaller or larger. The drawing is, therefore, scaled down or scaled up from the actual part.

The scale of a drawing may be any ratio, but the most common ones are 1:1, 1:2, 1:5, 1:10, etc. Scale may be shown in several different ways, such as FULL, 1 = 1, 1:1, 1/1, or some other similar form. Although many prints are made with all views drawn at the same scale, prints may also contain views drawn at different scales. Scale is indicated on the title block and applies to all views in the drawing unless otherwise indicated.

The image below shows drawings of a 1-inch cube at various scales.

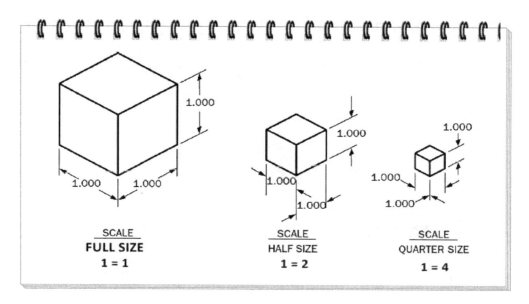

Some prints may show an axonometric view at actual size. The view makes it easier to visualize the size of the actual part.

The groove has been shown in a detail view at a larger scale to make it easier to read. The scale for the view is indicated next to the view.

The next image shows drawings of a 6-millimeter pyramid at various scales. Some prints may show an axonometric view at actual size.

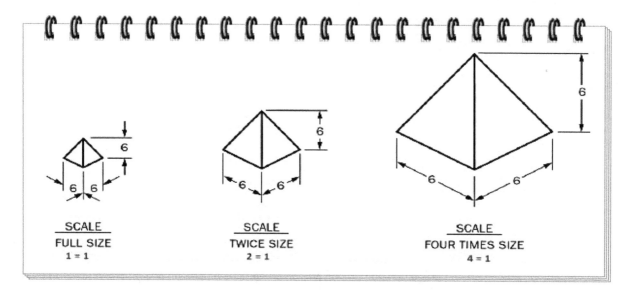

A line under a dimension indicates the dimension is not to scale; see the following images.

The line may be straight, or it may be wavy. As a rule, prints should never be scaled or measured to find a dimension; some are even marked with a note "DO NOT SCALE." Scaling prints may lead to serious problems for several reasons.

The drawing may not be drawn to the correct scale, or the dimension may have been changed. This is less of a concern with CAD-generated drawings than with hand-drawn prints where a complete redraw due to a small design change was labor-intensive and could be skipped. Regardless of how the print was generated, the paper may stretch or shrink.

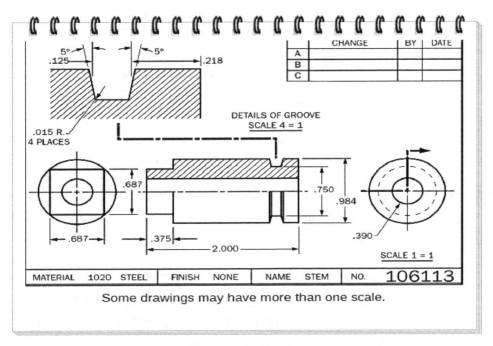
Some drawings may have more than one scale.

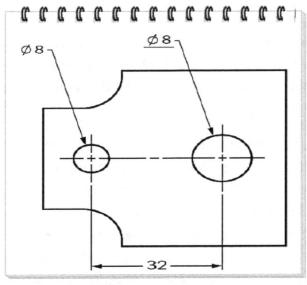

Angular dimensions are also are included on drawings. These dimensions are expressed in terms of degrees °, minutes ' and seconds"

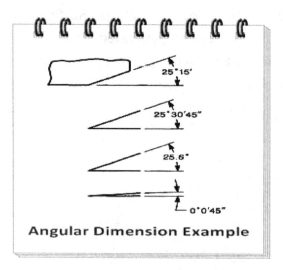

Angular Dimension Example

These dimensions are expressed by the symbols ° for degrees, ' for minutes and " for seconds. When degrees are indicated alone, the numerical value is followed by the symbol. If only minutes and seconds are used, the number of minutes and seconds is proceeded by 0° or 0°0'

The size of an object is indicated by numbers placed within dimension lines on a drawing. Dimension lines are solid, light, thin lines terminated with arrowheads. They are located between extension lines and placed on the drawing in a way that avoids confusion. This means the lines usually are placed beside rather than on the illustration of the object. Extension lines do not touch the object but rather are light lines that extend from the view of the object and bound the dimensions. Arrowhead points of the dimension lines touch the extension lines. Below are a set of general guidelines for draftspersons regarding the conventions for expressing dimensions as set forth by the American Standards' Association, they are as. follows:

1. Dimension lines are solid, light lines terminated by arrowheads.2. Arrowhead points touch the edge of the 'extension lines and usually are but 1/2" long.3. The dimension line is broken by a space in which the dimension is written. Extension lines are drawn at right angles to the location on the drawing to which they apply. Dimensions usually are placed so they can be read from the, bottom. Dimensions are grouped together and arranged so that they produce an orderly appearance. Numerals are staggered, not less than 1/4" apart to avoid confusion. Fractions within dimensions usually are larger than whole numbers with each number about,

two-thirds the height of a whole number. Do not repeat dimensions on the drawing; Do not originate lines so that they will cross either extension or other dimension lines. Do not originate or end dimensions on hidden lines. The most important dimensions should. be located with the principal or most important view of object. Most of these guidelines are illustrated in Dimensions provide information about both size and location. Length and width generally are the size directions. Location dimensions indicate where fasteners, holes, notches, arcs and so forth are located.

DIMENSIONS and DIMENSION LINES:

Dimensions complete the description or definition of size of a working drawing. Dimensions are indicated in inches, feet and inches, or decimals? The symbol for an inch is " while the symbol for afoot is '. It is standard practice to place a hyphen between feet and inches if both units are used on a drawing. For example, nine and one-half feet would be expressed as 9'-6". Usually if all measurements are in inches, the inch symbol is omitted, when the length exceeds 72", foot dimensions and marks are used.

Frequently in manufacturing industries, the decimal system of dimensions is used. It provides for exact control of manufactured parts. All figures on the drawing are shown as decimals. The standard convention is to measure and indicate two-place decimals to the right of the decimal point with the second number always being an even number.16, .82, .74rather than an odd number like, .67 or .05. The only exception to this convention occurs when greater precision in the part is required. In such instances, 3-place and 4-place dimensions are used. A fraction-decimal conversion chart is found in the Appendix of this booklet for your future reference.

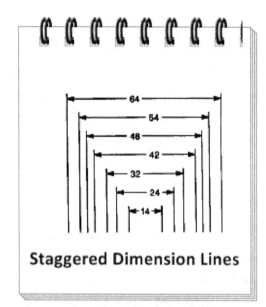
Staggered Dimension Lines

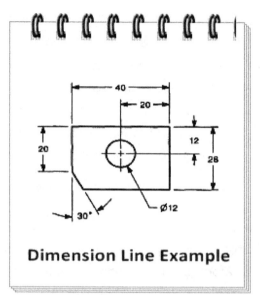
Dimension Line Example

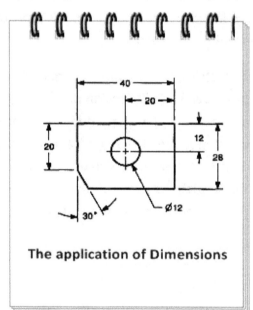
The application of Dimensions

An example of dimensioned round holes follows. Note that where it is not clear whether or not a hole goes through the part, the abbreviation THRU is placed after the dimension. On a blind hole the depth dimension is the full diameter from the outside surface of the workpiece. When the depth dimension is unclear, as an example when a depth comes from a curved surface, the depth is always dimensioned.

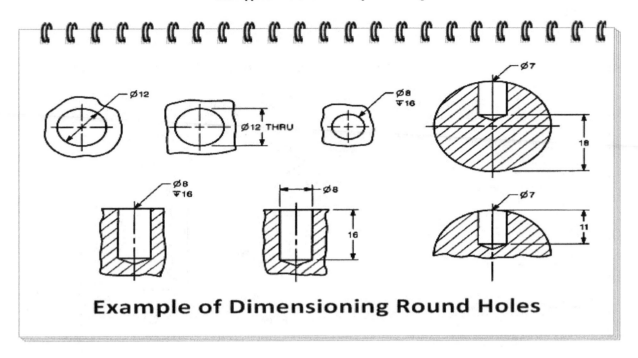

Example of Dimensioning Round Holes

CHAMFERS:

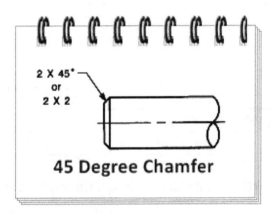

45 Degree Chamfer

When the edge of a round hole is chamfered, it is accepted practice to use a note to specify the degree of the chamfer, except where the diameter of the chamfer requires dimensional control. This kind of control can be applied to the diameter of the chamfer on a shaft. Important: A note is only used with a 45° Chamfer.

RADII:

A radius is any straight line extending from the center to the periphery of a circle or sphere.

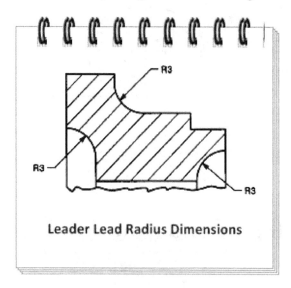

Leader Lead Radius Dimensions

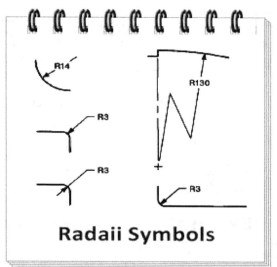

Radaii Symbols

When a leader line is directed to an arc or circle, its direction should always be radial. Each radius is proceeded by the appropriate radius symbol.

RADIUS	R
SPHERICAL RADIUS	SR
CONTROLLED RADIUS	CR

The radius dimension is sometimes extended through the radius center when space is limited.

Radius Tolerance:

The radius symbol R creates a zone that is defined by two arcs, the minimum and maximum radius. Between these two arcs lies the tolerance zone. A good part will lie between these two arcs, as seen on the next image.

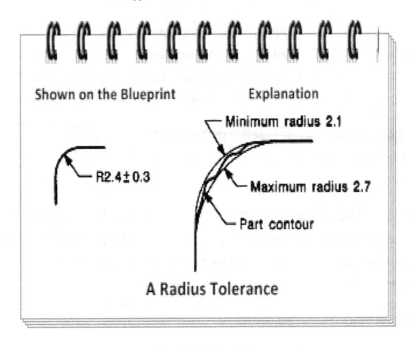

A Radius Tolerance

Controlled Radius Tolerance:

The controlled radius tolerance is defined by the symbol CR, again this tolerance zone is defined by the minimum and maximum radii. Remember the part surface must lie inside this tolerance zone.

Center of Radius:

When a dimension is given to the center of a radius, a small cross is drawn at the center of the radius. The drawing will show that the arc location is controlled by other features.

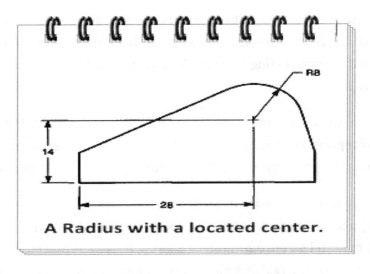

A Radius with a located center.

THE LANGUAGE OF BLUE PRINTS:

Prints may include English or non-English notes. Modern prints are made with the intent of maximizing readability by anyone within a global audience. The language used is likely to consist of concise statements using the simplest words and phrases for conveying the intended meaning. Certain words and phrases are frequently used on a print, such as:

- "PER," "CONFORMING TO," "AS SPECIFIED IN," and "IN ACCORDANCE WITH" (or "IAW") indicate that another document or standard is included in the requirements specified by the print, by reference.

- "UNLESS OTHERWISE SPECIFIED" (or "UOS") indicates a default requirement. This phrase can be found at the beginning of a note, and indicates that the default is a generally applied requirement. And the exception can be clarified by providing a reference to another document or requirement on the print.

- "SHALL" indicates a mandatory requirement.

- "WILL" indicates a declaration of purpose on the part of the design activity.

- "SHOULD" indicates a recommended practice.

- "MAY" indicates an allowed practice.

LEGENDS AND SYMBOLS:

A legend, if used, is placed in the upper right corner of a blueprint below the revision block. The legend explains or defines a symbol or special mark placed on the blueprint.

While lines are used to indicate the shape of objects, symbols are used to indicate both the location and types of items or materials. For example, symbols are used to indicate the type of material to be used in constructing objects, the location of, types of fasteners, fixtures and the types of equipment.

One example is an electrical diagram, symbols can indicate an outlet, switch or box location. On a construction blueprint, they indicate type of material to be used in parts of the structure while in installation and repair manuals they represent types of switches and valves.

Generally, drafting symbols are used instead of text notes for the sake of international readability. Reading a print requires understanding the symbols currently in use as well as prior practices that may still be encountered. A legend, if used, is placed in the upper right corner of a blueprint below the revision block. The legend explains or defines a symbol or special mark placed on the blueprint.

	TYPE OF TOLERANCE	CHARACTERISTIC	SYMBOL
FOR INDIVIDUAL FEATURES	FORM	STRAIGHTNESS	—
		FLATNESS	▱
		CIRCULARITY (ROUNDNESS)	○
		CYLINDRICITY	⌭
FOR INDIVIDUAL OR RELATED FEATURES	PROFILE	PROFILE OF A LINE	⌒
		PROFILE OF A SURFACE	⌓
FOR RELATED FEATURES	ORIENTATION	ANGULARITY	∠
		PERPENDICULARITY	⊥
		PARALLELISM	∥
	LOCATION	POSITION	⌖
		CONCENTRICITY	◎
		SYMMETRY	⌯
	RUNOUT	CIRCULAR RUNOUT	↗
		TOTAL RUNOUT	↗↗

Some Geometric Characteristic Symbols are shown in the above image. From time to time you might see a note accompanying a geometric symbol or a note in place of a geometric symbol. This is normally done when a symbol is not enough to convey geometric requirements.

CHAPTER 2

The Alphabet of Lines:

Just as each letter in our alphabet provides us with the ability to communicate our thoughts and ideas with each other. The lines used in the "Alphabet of Lines" also allow us to communicate thoughts and ideas. The most obvious reason for a line appearing in a drawing is to define the shape of an object. Lines are used for many other purposes, and the ability to recognize the type and purpose of a line in a drawing is the first step to understanding blueprints.

Being able to interpret lines, symbols, dimensions, notes, and other information on a print is a critical skill, and necessary for the manufacture of quality parts. I will try to provide you with definitions and examples of most common symbols used on prints today.

Visible lines represent features that can be seen in the current view. *Hidden lines* represent features that cannot be seen in the current view; a *Centerline* represents symmetry, path of motion, centers of circles, axis of symmetrical parts. *Dimension and Extension lines* indicate the sizes and location of features on a drawing.

To read blueprints, you must understand the use of lines. The alphabet of lines is the common language of the technician and the engineer. In drawing an object, a draftsman arranges the different views in a certain way, and then uses different types of lines to convey the information to the machinist.

Several different types of lines are used to make prints, each with its own purpose. Object lines indicate the visible edges of parts, while hidden lines indicate edges that are hidden behind the object. Dimensions, annotations, and notes are applied using their own special purpose lines.

Object Lines:

Object lines are medium, solid lines used for all lines on the drawing representing visible edges on the object, as seen below. Object lines are also called "outlines" or "visible lines."

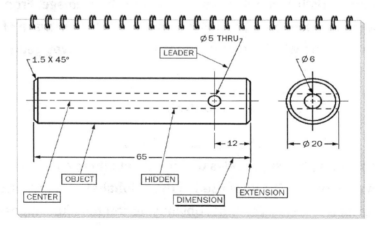

Hidden Lines:

- Hidden lines are thin, evenly spaced dashed lines used to show object edges that are not visible.

Center Lines:

- Center lines are alternate long/short dashes and are used to show the axes of rotation for axisymmetric objects.

Symmetry Lines:

• Symmetry lines are center lines with the addition of a pair of short thick continuous lines at each end, used to show the center of symmetrical objects.

Extension Lines:

Extension lines are thin, solid lines extending out from the object lines or center lines. They begin with a short, visible gap from the object line of the part and extend perpendicular to their associated dimension line. Extension lines may also be referred to as "witness lines" when they connect two collin-ear object lines to provide clarity that they are in fact collinear. They begin and end with short, visible gaps from the object lines of the part they connect.

Extension lines may also be referred to as intersection lines when they are used to show points or intersections. Intersecting extension lines extend slightly beyond their point of

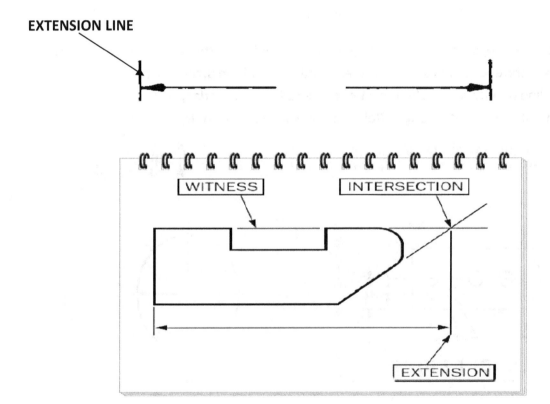

Dimension Lines:

• Dimension lines are thin, solid lines indicating the dimensions of the object. They connect extension lines and usually terminate in arrowheads, although other terminations are also used. Dimension lines may be broken to allow space for the dimension and tolerance, or they may underline the dimension and tolerance, depending on the drafting standard used.

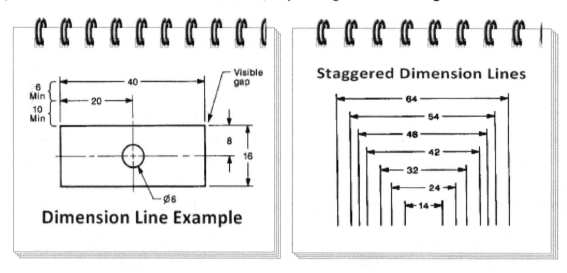

Leader Lines:

• Leader lines are thin, solid lines indicating a point or region on the part to which a number, note, or other reference applies. Leader lines terminate with an arrowhead when terminating on an object line or with a dot when terminating within the boundary of an object. Leader lines are generally drawn at an angle. Below are a few examples of leader lines.

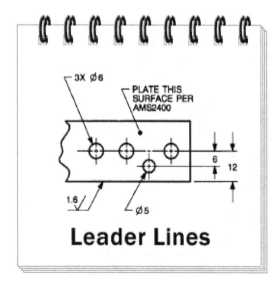
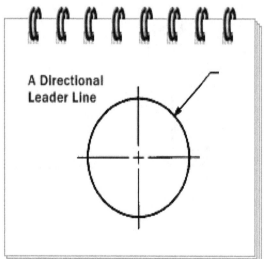

Break Lines:

- Break lines are thin, solid lines indicating a break in the view or a partial section view. Some objects have disproportionate dimensions, making them difficult to place on the print at an appropriate scale. It is then necessary to "break" them to present a reasonable level of detail. As the image below shows, there are three types of break lines commonly used:
- S-breaks: For round objects
- Z-breaks: For thin, long, wide objects
- Freehand breaks: For long, rectangular objects

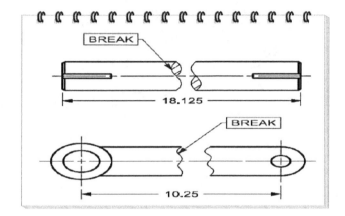

Sometimes long parts are drawn with break lines when the removed portions lack detail that is needed for print interpretation. The limits of the parts retained are shown as for partial views, and the portions appear close to each other.

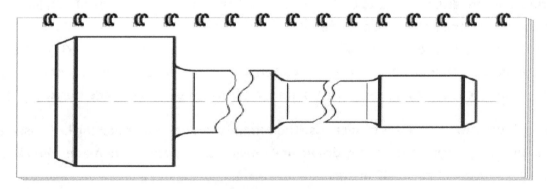

(Break Line Example)

Cutting Plane Lines:

Cutting plane lines are thick and consist of a long dash and two short dashes in a repeating pattern. Sometimes it is easier to visualize a part if it is drawn as though it was cut and a part of the object removed. A cutting plane line is used for this purpose. The cutting plane line indicates an imaginary cut, shown in the next image. The cut surface is shown in a separate view that is called a section view.

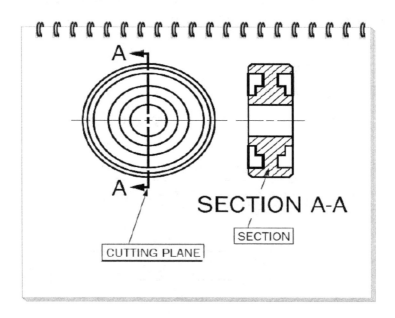

Section Lines:

- Section lines are also referred to as "crosshatching" and consist of a series of thin lines filling a region on the object, used to identify the cut surface in a section view. Crosshatching usually consists of a set of solid parallel lines at a 45-degree angle; however, section lines may consist of lines at other angles and in other patterns. If more than one component is sectioned, different crosshatching patterns are used for each part to provide clarity.

Hatching is used to show areas of sections. The simplest form of crosshatching lines used are continuous, thin, single lines, usually drawn at 45 degrees. Exceptions are made when the object lines are at or close to 45 degrees.

Separate areas of a section or separate sections of the same component are always hatched in an identical manner. The hatching of adjacent components is done with different line angles or spacing's. The figure below shows the hatching of four adjacent components.

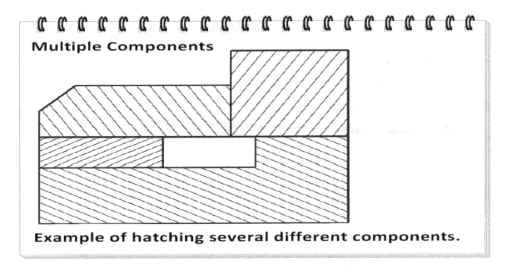

Example of hatching several different components.

Outline hatching can be used to Indicate the Type of Material. Sometimes different types of lines are shown as section lines for the purpose of referencing different materials.

IMPORTANT: Regardless of what types of lines are shown, the lines are never used as material specification; the type of material is <u>always</u> specified in a title block or separate note.

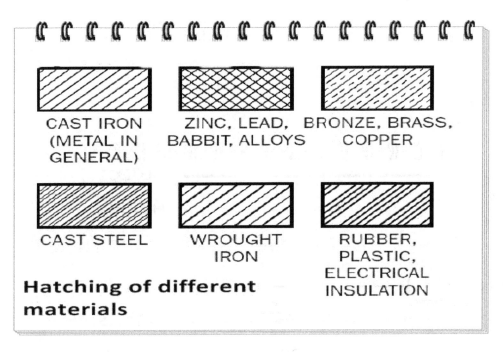

Hatching of different materials

PHANTOM LINES:

Phantom lines are thin and consist of a long dash and two short dashes in a repeating pattern. Phantom lines are used to indicate alternate positions of moving parts, repeated details, and material on a part before machining. Phantom lines indicate an alternative position or repeated details. Note how the Phantom lines indicate the engaged position of the quill feed lever in the image below.

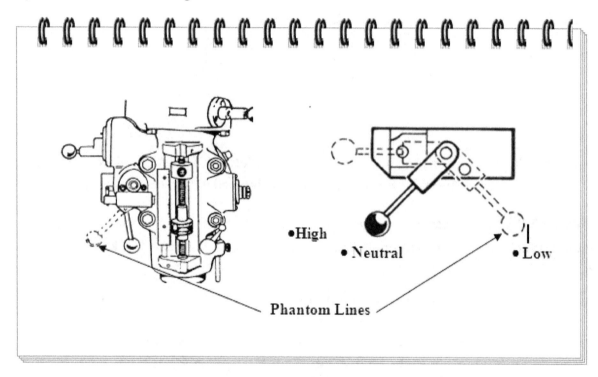

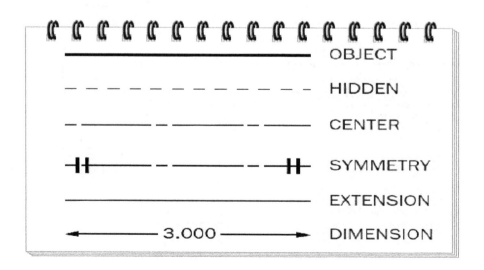

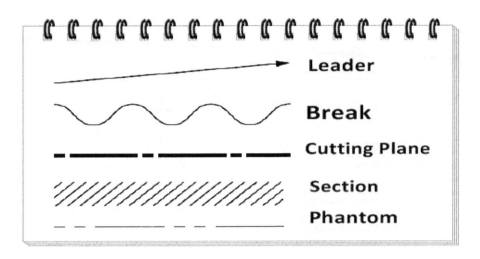

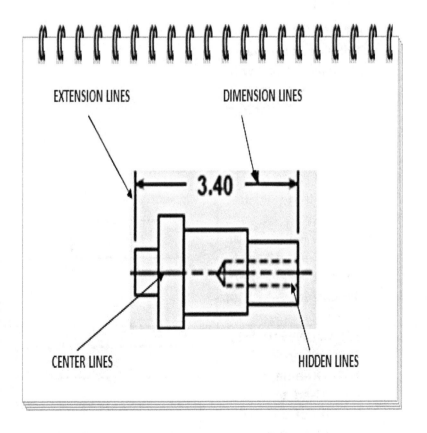

	OBJECT LINES Heavy unbroken lines, used to indicate visible edges. Also called Visible Lines.	
	HIDDEN LINES Medium lines, with short evenly spaced dashes. Used to indicate hidden edges.	
	EXTENSION LINES Thin unbroken lines, used to indicate the extent of dimensions.	
	DIMENSION LINES Thin lines terminated at each end with arrowheads. Used to indicate distance measured.	
	LEADER LINES A thin line terminated at one end with an arrowhead. Used to indicate a part, dimension or reference.	1/4 x 20 UNC-28 THD.
	BREAK LINES (LONG) Thin solid ruled lines with free-hand zig-zags. Used to reduce the size of a drawing and reduce detail.	

⌇	**BREAK LINES (SHORT)** Thick solid freehand lines used to indicate short breaks.	
┆	**DATUM LINE** A medium series of dashes one long and two short. Evenly spaced and ending with a long dash. Used to indicate an alternate position of parts or to indicate a datum line.	
┊	**STITCH LINE** A medium line of short dashes evenly spaced and labeled. Used to indicate stitching or sewing.	
↑⋯↑ ↑――↑	**CUTTING PLANE LINE** Used to indicate where an imaginary cut took place. **VIEWING PLANE LINE** Used to indicate direction of sight when a partial view is used.	
▨	**SECTION LINES** Used to indicate the surface in the section view imagined to have been cut along the cutting plane line.	

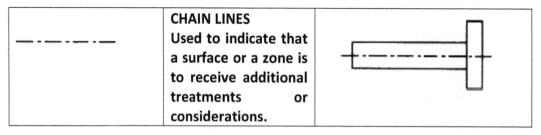

| | CHAIN LINES Used to indicate that a surface or a zone is to receive additional treatments or considerations. | |

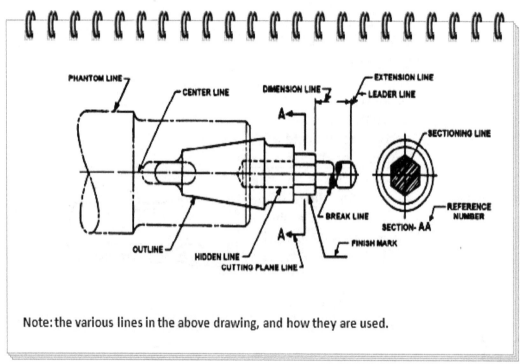

Note: the various lines in the above drawing, and how they are used.

CENTER LINES:

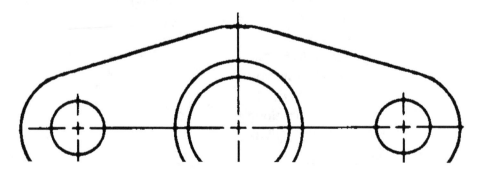

Center Lines consist of alternating long and short dashes (above fig). Use them to represent the axis of symmetrical parts and features, bolt circles, and paths of motion. You may vary the long dashes of the center lines in length, depending upon the size of the drawing. Start and end centerlines with long dashes and do not let them intersect at the spaces between dashes. Extend them uniformly and distinctly a short distance beyond the object or feature of the drawing unless a longer extension line is required for dimensioning or for some other purpose. Do not terminate them at other lines of the drawing, nor extend them through the space between views. Very short centerlines may be unbroken if there is no confusion with other lines.

Dimensioning and adding annotations are very important features in any drawing. If the dimension of any object could be drawn just by pointing the object, it will be the best facility. Such a feature is known as automatic dimensioning, particularly if the dimensions are drawn in the various views of the object. Generally, linear dimensioning involves the drawing of two extension lines, (a dimension line with arrow heads) separated from the object. Additionally tolerance values may also have to be shown in some cases. It can thus be realized that dimensioning is rather a complex process and many decisions are involved.

Choice should be available to draw the extension lines, in terms of location, length and the distance from the object. The location choice is generally by picking the end points of the object to be dimensioned.

NOTES AND SPECIFICATIONS:

Blueprints show all of the information about an object or part graphically. However, supervisors, contractors, manufacturers, and shop personnel may need more information that is not adaptable to the graphic form of presentation. Such information is shown on the drawings as notes or as a set of specifications attached to the drawings.

NOTES are placed on drawings to give additional information to clarify the object on the blueprint. Leader lines show the precise part notated.

A SPECIFICATION is a statement or document containing a description such as the terms of a contract or details of an object or objects not shown on a blue print or drawing. Specifications describe items so they can be manufactured, assembled, and maintained according to their performance requirements.

They furnish enough information to show that the item conforms to the description and that it can be made without the need for research, development, design engineering, or other help from the preparing organization.

CHAPTER 3
MULTI-VIEW DRAWINGS:

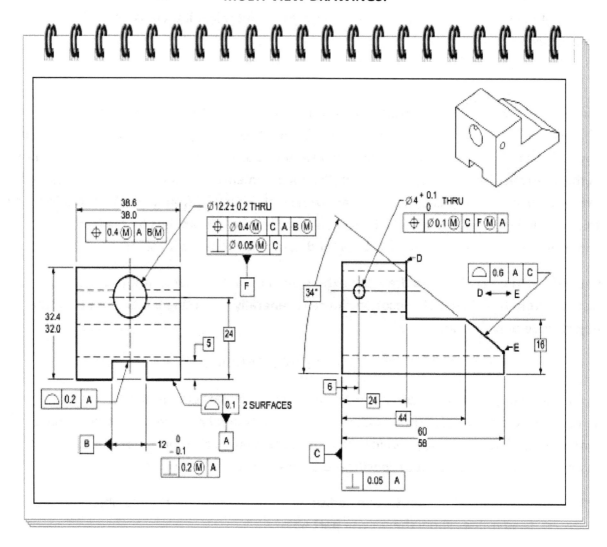

DETAIL DRAWINGS:

A detail view is indicated using one of the notations shown below. The detail view itself is labeled with the same letter(s) and scale. The word "DETAIL" may be shown with the view label. If the detail view is **found on a different sheet from that of its parent, references will be included.**

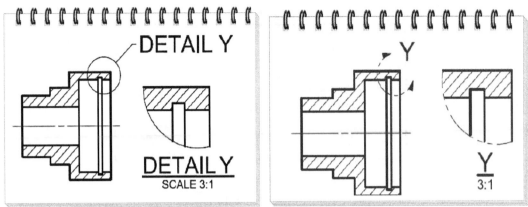

(Detailed views of Symmetrical Parts)

Symmetry lines indicate that the view in the print shows only a portion of the part. This is not common with **CAD** prints, where the view creation is done automatically, but it is still used as a way to save print space.

The complexity of the shape of a drawing governs the number of views needed to project the drawing. Complex drawings normally have six views: two ends, a front, a top, a rear, and a bottom. However, most drawings are less complex and are shown in three views. We will explain both in the following paragraphs.

In learning to read blueprints, you must develop the ability to visualize the object to be made from the blueprint. You cannot read a blueprint all at once any more than you can read an entire page of print all at once. When you look at a multi-view drawing, first survey all of the views, and then select one view at a time for more careful study. Look at adjacent views to determine what each line represents.

Each line in a view represents a change in the direction of a surface, but you must look at another view to determine what the change is. A circle on one view may mean either a hole or a protruding boss. When you look at the top view you see two circles, and you must study the other view to understand what each represents. A glance at the front view shows that the smaller circle represents a hole (shown in dashed lines), while the larger circle represents a protruding boss. In the same way, you must look at the top view to see the shape of the hole and the protruding boss.

You can see from the example that you cannot read a blueprint by looking at a single view, if more than one view is shown. Sometimes two views may not be enough to describe an object; and when there are three views, you must view all three to be sure you read the shape correctly.

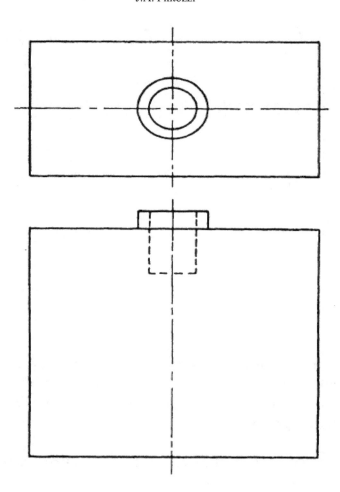

PROJECTIONS:
In blueprint reading, a view of an object is known technically as a projection. Projection is done, in theory, by extending lines of sight called projectors from the eye of the observer through lines and points on the object to the plane of projection.

Drawings are two-dimensional representations of three-dimensional parts to be built. Creating a two-dimensional representation of a three dimensional part requires the draftsman to project the part onto a two-dimensional surface. There are many methods of projection used for various purposes. The process of projection can be illustrated using four concepts: the object being projected, the plane of projection, projection lines, and the observer's viewpoint.

Parallel projection uses an observer's viewpoint located an infinite distance from the object. Stated differently, all projection lines are parallel to one another. This has the effect that a line on the object is projected into a line on the projection plane. Parallel lines on the object become parallel on the projection. The midpoint of a line on the object is projected to the midpoint of the line of the projection.

There are several types of parallel projections, each following slightly different rules. The main types are orthographic projection and oblique projection. In oblique projection, the projection lines cross the projection plane at a non-perpendicular angle. Oblique projection includes the subtypes cabinet, cavalier, and military. Oblique projections were common as recently as the nineteenth century in technical, architectural, and military depictions. Modern blueprints use orthographic, not oblique, projection.

ISOMETRIC PROJECTION:

Isometric projection is the most frequently used type of axonometric projection, which is a method used to show an object in all three dimensions in a single view. Axonometric projection is a form of orthographic projection in which the projectors are always perpendicular to the plane of projection. However, the object itself, rather than the projectors, are at an angle to the plane of projection. The figure below shows a cube projected by isometric projection.

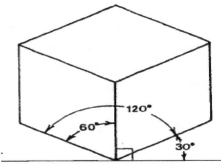

Isometric drawings show three surfaces of the selected object, each_ on a different axis. One axis is vertical; the other two are drawn to the right and left at an angle of 30 ° to the horizontal so that the object can be rotated right or left about the vertical axis. circles drawn in the isometric form are ellipses.

Note: The cube is angled so that all of its surfaces make the same angle with the plane of projection. As a result, the length of each of the edges shown in the projection is somewhat shorter than the actual length of the edge on the object itself. This reduction is called foreshortening. Since all of the surfaces make the angle with the plane of projection, the edges foreshorten in the same ratio. Therefore, one scale can be used for the entire layout; hence, the term *isometric* which literally means the same scale.

ORTHOGRAPHIC PROJECTION:

Working drawings must show a great deal of detail and information. Therefore, actual pictures are not suitable. They are too cluttered and confusing. Instead, the most frequently used type of drawing is the orthographic projection. The orthographic projection is a multi-view drawing of an object where each view is at right angles to every other view. The views are projected on flat surfaces and do not show perspective.

To illustrate what an orthographic projection is, imagine an object, in a clear plastic box. Further, imagine that each side of the plastic box is on hinges so that it opens out flat on a surface as illustrated below. Now, looking straight into each side, including top and bottom, of the box, imagine that you trace the outline of the object in the box on the plastic surfaces. Then you open the box out flat on a surface. Notice that the box and object within the box has six views: top, bottom, front, and back, left side and right side. Each side is a possible view to include in an orthographic projection. However, most orthographic projections show only three of the six views front, top and right side. The front is usually the most critical. However, you should select the three views that offer the best description of the object in question and show the object in its natural position. The Figure Below is an example of an orthographic projection.

The figure below shows an object placed in a transparent box hinged at the edges. With the outlines scribed on each surface and the box opened and laid flat as shown below. The result is a six-view orthographic projection. The rear plane is hinged to the right side plane, but it could hinge to either of the side planes or to the top or bottom plane. View B shows that the projections on the sides of the box are the views you will see by looking straight at the object through each side. Most drawings will be shown in three views, but occasionally you will see two-view drawings, particularly those of cylindrical views.

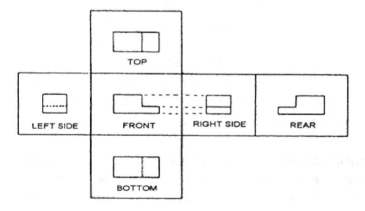

A THREE-VIEW Orthographic Projection drawing shows the front, top, and right sides of an object.

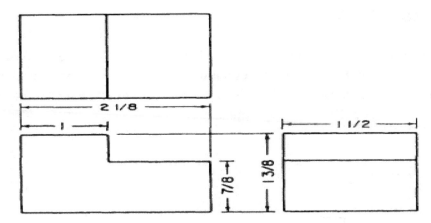

The views are usually in the positions shown above. The front view is always the starting point and the other two views are projected from it. You may use any view as your front view as long as you place it in the lower-left position in the three-view. This front view was selected because it shows the most characteristic feature of the object, the notch. The right side or end view is always projected to the right of the front view. Note that all horizontal outlines of the front view are extended horizontally to make up the side view. The top view is always projected directly above the front view and the vertical outlines of the front view are extended vertically to the top view.

To clarify the three-view drawing further, think of the object as immovable and visualize yourself moving around it. This will help you relate the blueprint views to the physical appearance of the object. Now study the next three-view drawing shown below Note: It is similar to the figure above with one exception; the object in the next figure has a hole drilled in its notched portion. The hole is visible in the top view, but not in the front and side views.

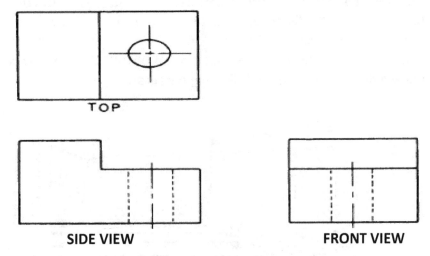

Therefore, hidden (dotted) lines are used in the front and side views to show the exact location of the walls of the hole.

The three-view drawing shown in the above figure, introduces two symbols that are not shown in the previous figure. They are hidden lines, that show lines that you normally cannot see on the object, and a center line that gives the location of the exact center of the drilled hole. The shape and size of the object are the same.

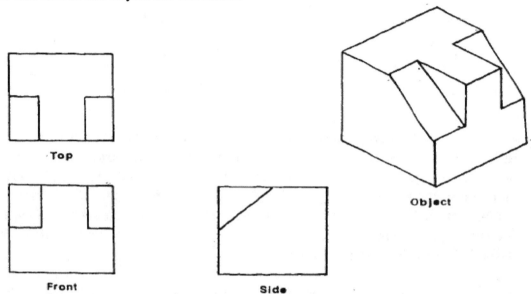

PERSPECTIVE DRAWINGS:

A perspective drawing is the most used method of presentation used in technical illustrations in the commercial and architectural fields. The drawn objects appear proportionately smaller with distance, as they do when you look at the real object. It is difficult to draw, and since the drawings are drawn in diminishing proportion to the edges represented, they cannot be used to manufacture an object. Other views are used to make objects and we will discuss them in the following paragraphs.

(An example of a perspective view)

SPECIAL VIEWS:

In many complex objects, it is often difficult to show true size and shapes orthographically. Therefore, the draftsmen must use other views to give engineers and craftsmen a clear picture of the object to be constructed. Among these are a number of special views, some of which we will discuss in the following paragraphs.

AUXILIARY VIEWS:

Auxiliary views are used in orthographic projections to show the true shape of features that are not aligned to the drawing's standard up-down and left-right projection directions. Objects that have features on slanted surfaces may be difficult to visualize without auxiliary views. Auxiliary views conform to all normal rules of orthographic projection, but their projected direction is whatever direction is needed to illustrate the part; see the image below.

Auxiliary views may also be partial views. Partial views are connected to the principal view by a center line or continuous thin line. Partial views are connected to the principal view by a center line or continuous thin line..

Auxiliary views are often necessary to show the true shape and length of inclined surfaces, or other features that are not parallel to the principal planes of projection. Look directly at the front view of the figure below. Notice the inclined surface. Now look at the right side and top views. The inclined surface appears foreshortened, not its true shape or size. In this case, the draftsman will use an auxiliary view to show the true shape and size of the inclined face of the object.

It is drawn by looking perpendicular to the inclined surface. The Figure below shows the principle of the auxiliary view. Look back to the figure, which shows an immovable object being viewed from the front, top, and side. Find the three orthographic views, and compare them.

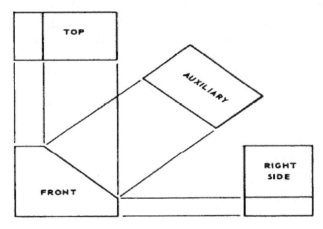

It should clearly explain the reading of the auxiliary view. (SEE ABOVE) for a comparison of orthographic and auxiliary views. View A shows a foreshortened orthographic view of an inclined or slanted surface whose true size and shape are unclear. View B uses an auxiliary projection to show the true size and shape. The projection of the auxiliary view is made by the observer moving around an immovable object, and the views are projected perpendicular to the lines of sight. Remember, the object has not been moved; only the position of the viewer has changed.

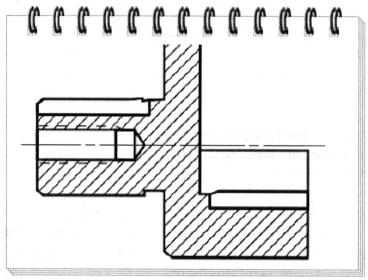

No cutting plane line Half Sections

A half section is similar to a full section except only half of one view is cut and removed; see above. The remaining half is shown as a basic view. Half **sections are commonly used for parts that are symmetrical.**

An offset section has two or more cuts that are parallel to each other but offset. An aligned section has two or more cutting planes that intersect but meet at an angle. The section view is projected to a location on the drawing from either of the angled cutting plane lines. The section view is drawn as though the two cuts were rotated into one view.

Sections can also be cut through parts using a combination of offset and aligned methods.

CHAPTER 4
SECTIONAL VIEW DRAWINGS:

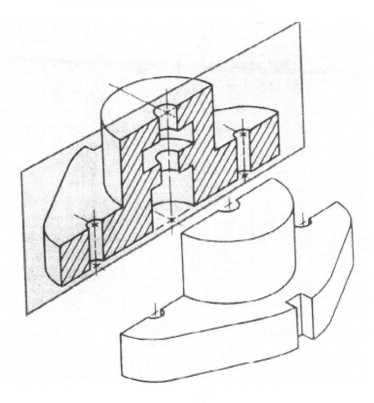

Section views give a clearer view of the interior or hidden features of an object that you normally cannot see clearly in other views. A section view is made by visually cutting away a part of an object to show the shape and construction at the cutting plane.

Notice the cutting plane line AA in the front view shown in figure 3-17, view A. It shows where the imaginary cut has been made. In view B, the isometric view helps you visualize the cutting plane. The arrows point in the direction in which you are to look at the sectional view.

View C is another front view showing how the object would look if it were cut in half. In view D, the orthographic section view of section A-A is placed on the drawing instead of the confusing front view in view A. Notice how much easier it is to read and understand.

When sectional views are drawn, the part that is cut by the cutting plane is marked with diagonal (or crosshatched), parallel section lines. When two or more parts are shown in one view, each part is sectioned or crosshatched with a different slant. Section views are necessary for a clear understanding of complicated parts. On simple drawings, a section view may serve the purpose of additional views.

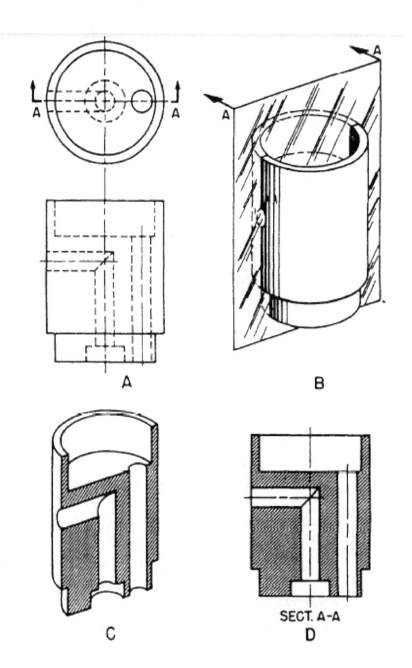

Section A-A in view D is known as a full section because the object is cut completely through.

OFFSET SECTION:

In this type of section, the cutting plane changes direction backward and forward (zigzag) to pass through features that are important to show. The offset cutting plane in the figure below is positioned so that the hole on the right side will be shown in section. The sectional view is the front view, and the top view shows the offset cutting plane line.

An offset section has two or more cuts that are parallel to each other but offset, see the image below. An aligned section has two or more cutting planes that intersect but meet at an angle; see Figure 4-25. The section view is projected to a location on the drawing from either of the angled cutting plane lines. The section view is drawn as though the two cuts were rotated into one view. Sections can also be cut through parts using a combination of offset and aligned methods.

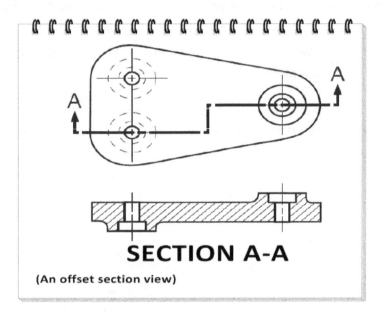

(An offset section view)

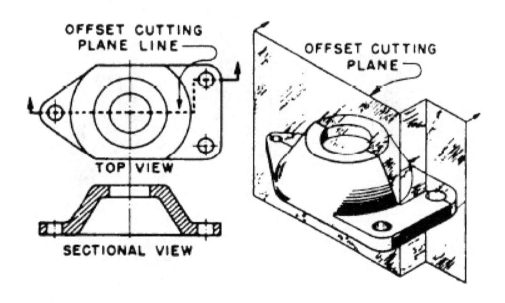

HALF SECTION:

It is used when an object is symmetrical in both outside and inside details. One-half of the objects are sectioned; the other half is shown as a standard view. The object shown is cylindrical and cut into two equal parts. Those parts are then divided equally to give you four quarters. Now remove a quarter. This is what the cutting plane has done in the pictorial view; a quarter of the cylinder has been re- moved so you can look inside. If the cutting plane had extended along the diameter of the cylinder, you would have been looking at a full section. The cutting plane in this drawing extends the distance of the radius, or only half the distance of a full section, and is called a half section.

The arrow has been inserted to show your line of sight. What you see from that point is drawn as a half section in the orthographic view. The width of the orthographic view represents the diameter of the circle. One radius is shown as a half section, the other as an external view.

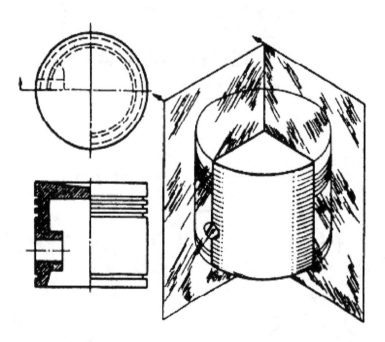

REVOLVED SECTIONS:

This type of section is used to eliminate the need to draw extra views of rolled shapes, ribs, and similar forms. It is really a drawing within a drawing, and it clearly describes the object's shape at a certain cross section. The draftsman has revolved the section view of the rib so you can look at it head on. Because of this revolving feature, this kind of section is called a revolved section.

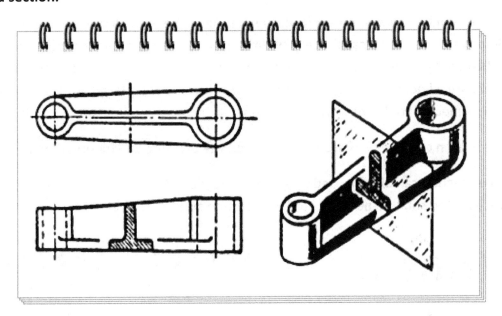

The appearance of a cross-section view superimposed on its parent view is a revolved section and indicates a cross-sectional view of the part, the cutting plane is indicated by the centerline of the cross section view. Revolved section views are used for symmetrical cross sections. Revolved section views can also be shown within a break.

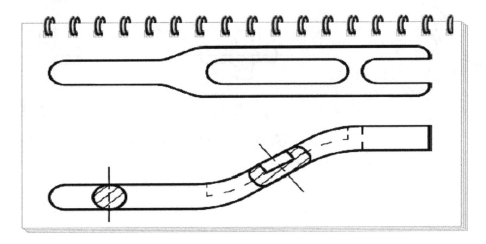

REMOVED SECTIONS:

This type of section is used to illustrate particular parts of an object. It is drawn like the revolved section, except it is placed at one side to bring out important details. It is often drawn to a larger scale than the view of the object from which it is removed.

A removed section is shown removed from the outline of the main drawing; see the images below. When the cross-sectional shape is symmetrical, the removed section may be drawn in line with the cutting plane, or it may be drawn in a different location and connected to the main object with a line. When removed, the section view may be placed: Near to the main object and connected to the view by a center line.

A removed section can also be placed anywhere on the drawing and identified with a label.

The removed section can also be used to enlarge a portion of an object. This method is used when an area is too small to be illustrated at the same scale as the rest of the object.

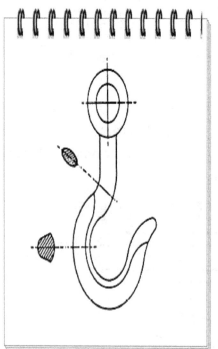

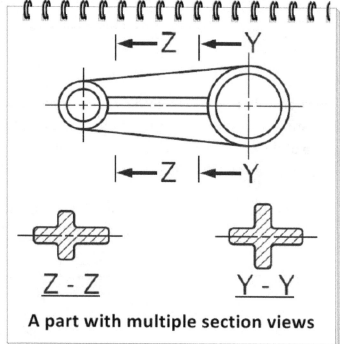

A part with multiple section views

ALLIGNED FEATURES:

If a part has nonsymmetrical features such as spokes, ribs, and webs, it is common practice to align the features in the drawing view. This is usually done with both regular views and section views. When the features are aligned in sectional views, the features are not shown in section. The section view below that is labeled "ALIGNED" has the lower spoke aligned or rotated into the section view.

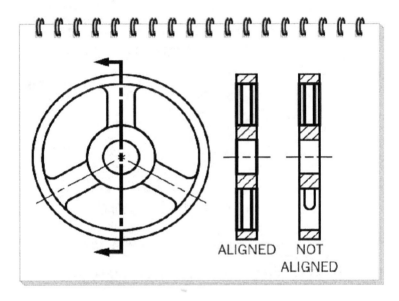

THE EXPLODED VIEW (Assembly Drawing):

 This is another type of view that is helpful and easy to read. The exploded view (shown below) is used to show the relative location of parts, and it is particularly helpful when you must assemble complex objects. Notice how parts are spaced out in line to show clearly each part's relationship to the other parts.

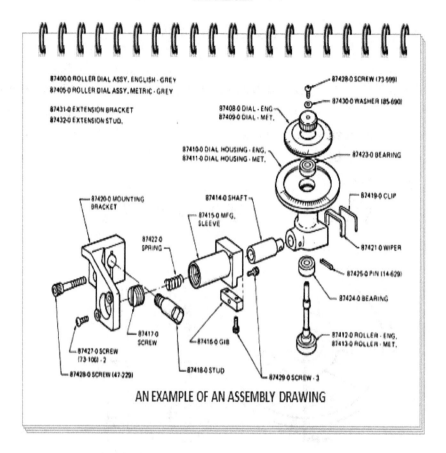

(AN EXAMPLE OF AN ASSEMBLY DRAWING)

DETAIL DRAWINGS REVISITED:

Remember that a detail drawing is a print that shows a single component or part. It includes a complete and exact description of the part's shape and dimensions, and how it is made. A complete detail drawing will show in a direct and simple manner the shape, exact size, type of material, finish for each part, tolerance, necessary shop operations, number of parts required, and so forth. A detail drawing is not the same as a detail view. A detail view shows part of a drawing in the same plane and in the same arrangement, but in greater detail to a larger scale than in the principal view. The appearance of a cross-section view superimposed on its parent view is a revolved section and indicates a cross-sectional view of the part, the cutting plane indicated by the cross-sectional view's center line.

Revolved section views are used for symmetrical cross sections. Revolved section views may also be shown within a break, as in Figure 4-31.

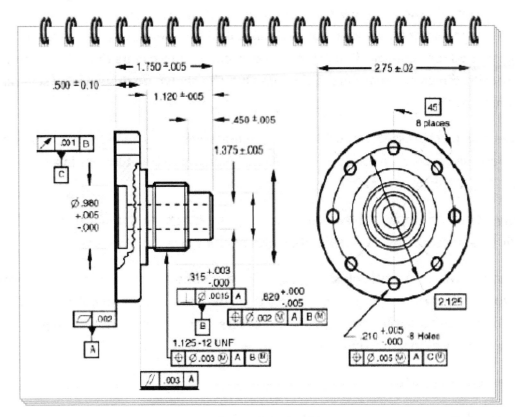

(EXAMPLE OF A DETAIL DRAWING)

The above image shows a common detail drawing. Study the figure closely and apply the principles for reading detail drawings given earlier. Notice that the dimensions on the detail drawing above are both conventional and examples of GD&T. The Figure below is an example of an isometric view.

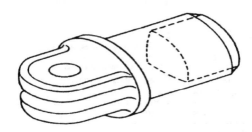

TOLERANCES:

Engineers realize that absolute accuracy is impossible, so they figure how much variation is permissible. This allowance is known as tolerance. It is stated on a drawing as (plus or minus) a certain amount, either by a fraction or decimal. Limits are the maximum and/or minimum values prescribed for a specific dimension, while tolerance represents the total amount by which a specific dimension may vary. Tolerances may be shown on drawings by several different methods;

Tolerances may be expressed: Implicitly—as a general tolerance applied to all dimensions unless otherwise specified. Implicit tolerances are given in the title block or in a supplemental document or CAD file referenced from the print.

• Explicitly—as direct limits or tolerance values applied directly to a dimension. As a geometric tolerance that may be associated with one or more basic dimensions.

All industrial prints have a title block that contains a set of tolerances, these tolerances are referred to as "general tolerances, They apply to all dimensions that do not have a direct tolerance or a geometric tolerance and are not covered by a separate note. The primary ways to indicate tolerances in a drawing are:

• A general tolerance note.

• A note providing a tolerance for a specific dimension.

• A reference on the drawing to another document that specifies the required tolerances.

• These are similar to specifying dimension units.

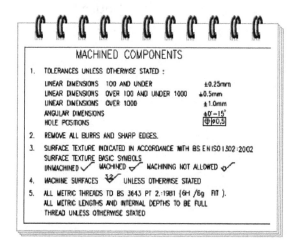

- The unilateral method (view A) is used when variation from the design size is permissible in one direction only.

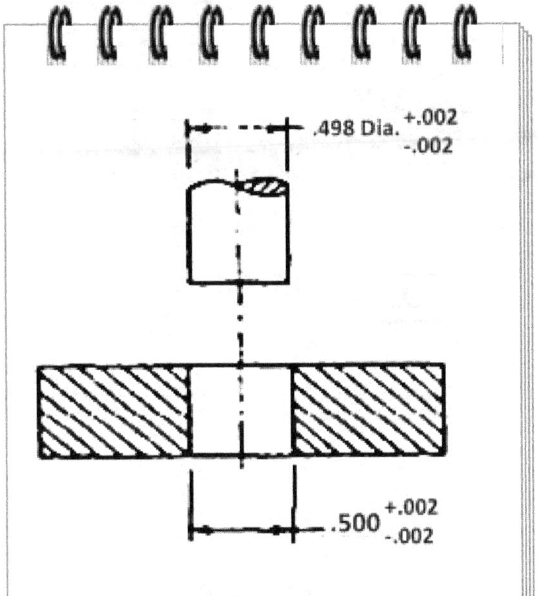

(View A)

- In the bilateral method, the dimension figure shows the plus or minus variation that is acceptable.

- In the limit dimensioning method, the maximum and minimum measurements are both stated.

The surfaces being toleranced have geometrical characteristics such as roundness, or perpendicularity to another surface. The illustration below shows typical geometrical characteristic symbols and how a feature control symbol may include datum references. A datum is a surface, line, or point from which a geometric position is to be determined or from which a distance is to be measured.

Any letter of the alphabet except I, O, and Q may be used as a datum identifying symbol. A feature control symbol is made of geometric symbols and tolerances.

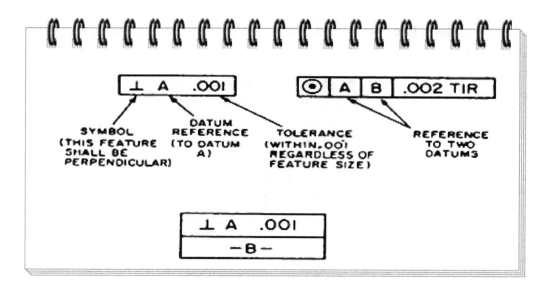

CHAPTER 5
THREADS, GEARS, WELDING SYMBOLS AND FINISH MARKS:

Looking at the first pair of charts below we see ten types of external threads. The next chart illustrates the parts of external threads. To save time, the draftsman uses symbols that are not drawn to scale. The drawing shows not only the dimensions of the threaded part, other information may be placed in "notes" almost any place on the drawing but most often in the upper left corner.

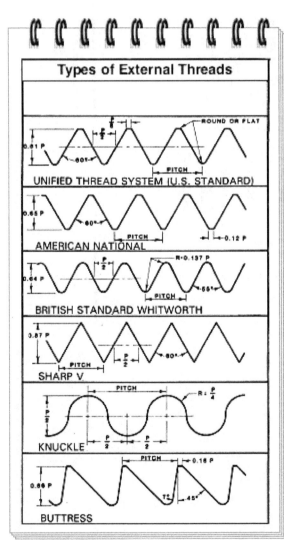
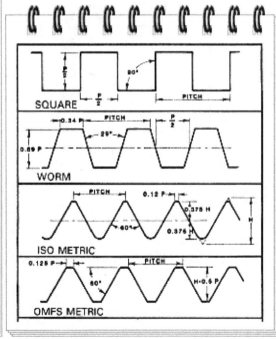

Parts of External Threads

Parts of a threaded fastener

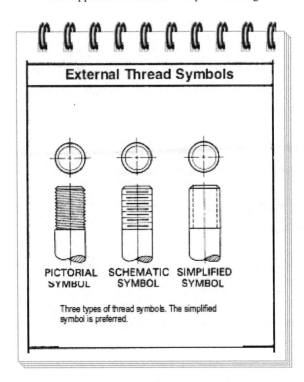

Three types of thread symbols. The simplified symbol is preferred.

When creating blueprints designers often choose the quickest drawing method to communicate their ideas and concepts. The illustrations shown above are all acceptable representations of threads, however the simplified symbol is usually chosen.

THE FIRST NUMBER OF THE NOTE, 1/4, IS THE NOMINAL SIZE, WHICH IS THE OUTSIDE DIAMETER. THE NUMBER AFTER THE FIRST DASH, 20, MEANS THERE ARE 20 THREADS PER INCH THE LETTERS UNC IDENTIFY THE THREAD SERIES AS UNIFIED NATIONAL COARSE. THE LAST NUMBER, 2, IDENTIFIES THE CLASS OF THREAD AND TOLERANCE, COMMONLY CALLED THE FIT.
If it is a left-hand thread, a dash and the letters LH will follow the class of thread. Threads without the LH are right-hand threads.

Specifications necessary for the manufacture of screws include thread diameter, number of threads per inch, thread series, and class of thread The two most widely used screw-thread series are (1) Unified or National Form Threads, which are called National Coarse, or NC, and (2) National Fine, or NF threads. The NF threads have more threads per inch of screw length than the NC.

Classes of threads are distinguished from each other by the amount of tolerance and/or allowance specified. Classes of thread were formerly known as *class of fit,* a term that will probably remain in use for many years.

The new term, *class of thread,* was established by the National Bureau of Standards in the *Screw-Thread Standards for Federal Services,* Handbook H-28.

Figure 4-13 shows the terminology used to describe screw threads. Each of the terms is explained in the following list:

• HELIX—The curve formed on any cylinder by a straight line in a plane that is wrapped around the cylinder with a forward progression.

• EXTERNAL THREAD—A thread on the outside of a member. An example is the thread of a bolt.

• INTERNAL THREAD—A thread on the inside of a member. An example is the thread inside a nut.

• MAJOR DIAMETER—The largest diameter of an external or internal thread

• AXIS—The center line running lengthwise through a screw.

• CREST—The surface of the thread corresponding to the major diameter of an external thread and the minor diameter of an internal thread.

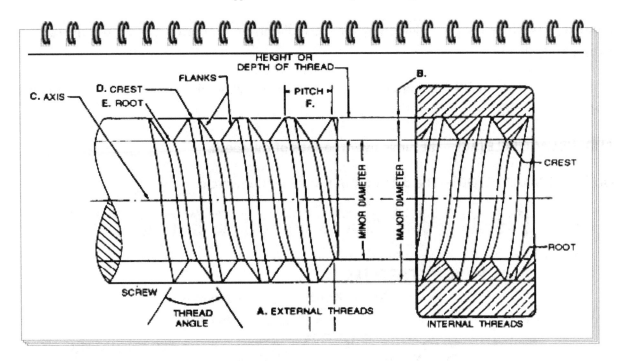

ROOT—The surface of the thread corresponding to the minor diameter of an external thread and the major diameter of an internal thread.

DEPTH—The distance from the root of a thread to the crest, measured perpendicularly to the axis.

PITCH—The distance from a point on a screw thread to a corresponding point on the next thread, measured parallel to the axis.

LEAD—The distance a screw thread advances on one turn, measured parallel to the axis. On a single-thread screw, the lead and the pitch are identical; on a double-thread, screw the lead is twice the pitch; on a triple-thread screw, the lead is three times the pitch.

GEARS:

When gears are drawn on machine drawings, the draftsman usually draws only enough gear teeth to identify the necessary dimensions.

PITCH DIAMETER (PD)—The diameter of the pitch circle (or line), which equals the number of teeth on the gear divided by the diametral pitch.

DIAMETRAL PITCH (DP)—The number of teeth to each inch of the pitch diameter or the number of teeth on the gear divided by the pitch diameter. Diametral pitch is usually referred to as simply PITCH.

NUMBER OF TEETH (N)—The diametral pitch multiplied by the diameter of the pitch circle (DP x PD).

ADDENDUM CIRCLE (AC)—The circle over the tops of the teeth.

OUTSIDE DIAMETER (OD)—The diameter of the addendum circle.

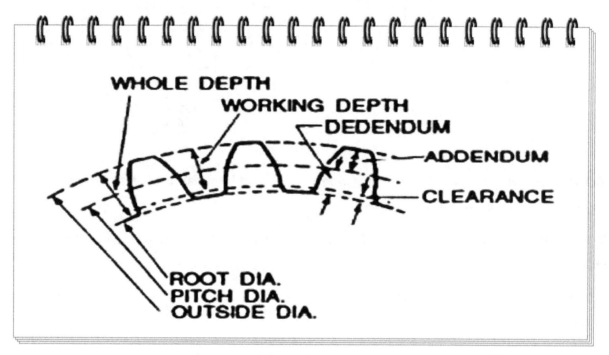

- **CIRCULAR PITCH (CP)**—The length of the arc of the pitch circle between the centers or corresponding points of adjacent teeth.

- **ADDENDUM (A)**—The height of the tooth above the pitch circle or the radial distance between the pitch circle and the top of the tooth.

- **DEDENDUM (D)**—The length of the portion of the tooth from the pitch circle to the base of the tooth.

- **CHORDAL PITCH**—The distance from center to center of teeth measured along a straight line or chord of the pitch circle.

- **ROOT DIAMETER (RD)**—The diameter of the circle at the root of the teeth.

- **CLEARANCE (C)**—The distance between the bottom of a tooth and the top of a mating tooth.

- **WHOLE DEPTH (WD)**—The distance from the top of the tooth to the bottom, including the clearance.

- **FACE**—The working surface of the tooth over the pitch line.

- **THICKNESS**—The width of the tooth, taken as a chord of the pitch circle.

- **PITCH CIRCLE**—The circle having the pitch diameter.

- **WORKING DEPTH**—The greatest depth to which a tooth of one gear extends into the tooth space of another gear.

- **RACK TEETH**—A rack may be compared to a spur gear that has been straightened out. The linear pitch of the rack teeth must equal the circular pitch of the mating.

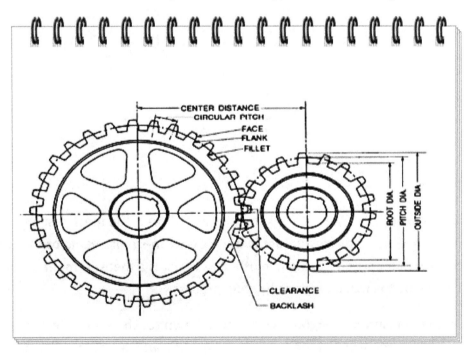

FINISH MARKS:

Finish marks (√) used on machine drawings show surfaces to be finished by machining. Machining provides a better surface appearance and a better fit with closely mated parts. Machined finishes are NOT the same as finishes of paint, enamel, grease, chromium plating, and similar coatings.

When working with the military for example, standards for finish marks are set forth in ANSI 46.1-1962. Many metal surfaces must be finished with machine tools for various reasons. The acceptable roughness of a surface depends upon how the part will be used. Sometimes only, certain surfaces of a part need to be finished while others are not. A modified symbol (check mark) with a number or numbers above it is used to show these surfaces and to specify the degree of finish.

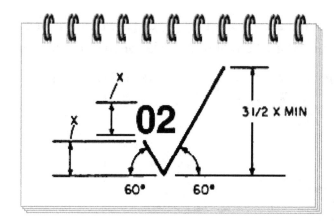

The number in the angle of the check mark, in this case 02, tells the machinist what degree of finish the surface should have. This number is the root-mean-square value of the surface roughness height in millionths of an inch. In other words, it is a measurement of the depth of the scratches made by the machining or abrading process.

Wherever possible, the surface roughness symbol is drawn touching the line representing the surface Figure 4-17.—Proportions for a basic finish symbol. which it refers. If space is limited, the symbol may be placed on an extension line on that surface or on the tail of a leader with an arrow touching that surface as shown in figure 4-18.

When a part is to be finished to the same roughness all over, a note on the drawing will include the direction "finish all over" along the finish mark and the proper number. An example is FINISH ALL OVER32. When a part is to be finished all over but a few surfaces vary in roughness, the surface roughness symbol number or numbers are applied to the lines representing these surfaces and a note on the drawing will include the surface roughness symbol for the rest of the surfaces.

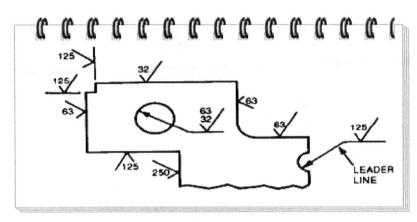

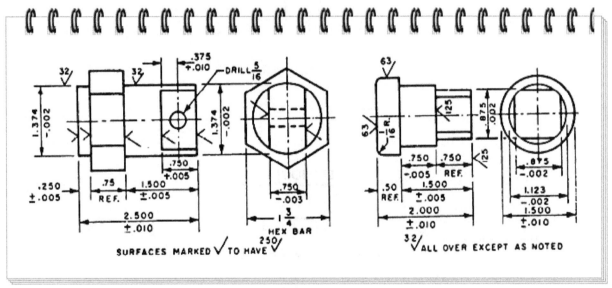

WELDING:

Machine Parts can be assembled and held in place using any of several methods, other than mechanical fasteners. Parts can be welded, soldered, or brazed, or they might be bonded or cemented with an adhesive. These methods are becoming popular and are replacing many threaded assemblies. Welding ranges from a simple spot weld, to one of the more complicated methods. Bonding may range from a simple solvent bonding of plastics, to fastening of materials with specially compounded adhesives. Procedures for fastening by welding or bonding are specified on prints with symbols and or separate notes.

There are five common basic types of weld joints that characterize the position of parts being welded: butt, tee, corner, lap, and edge.

TYPES OF JOINTS:

1. Butt Joint

2. Tee Joint

3. CORNER

4. LAP JOINT

5. EDGE JOINT

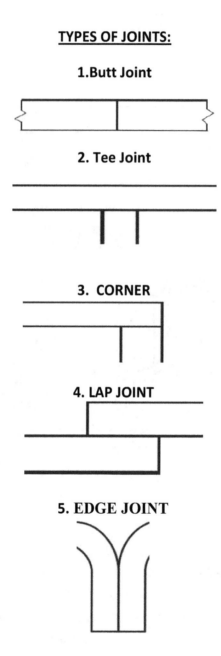

The Apprentices Guide to Blueprint Reading

0TYPES OF WELDS: GAS AND ARC

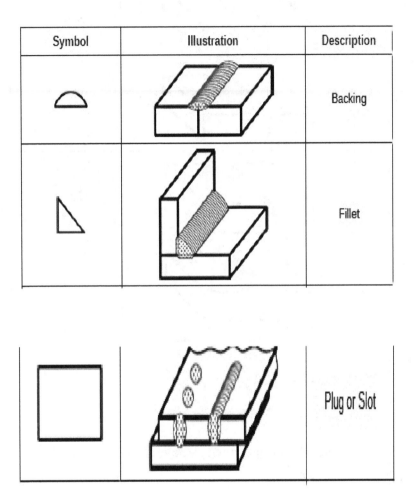

Plug and slot welds are typically used on lap joints, though in some cases they may also be applied to tee joints. Unlike other joints, a lap joint may be illustrated in a plan view or in the standard elevation view available for other joints. When the weld symbol is applied in the plan view, the "arrow side" is the member visible on top in that view; the "far side" is the hidden member. When applied in an elevation view, the "arrow side"-"far side" designation follows the convention for other joints shown in elevation view. A hole or slot is made in one member to receive the weld. The hole or slot is produced in the member indicated by the arrow-side designation of the basic weld symbol.

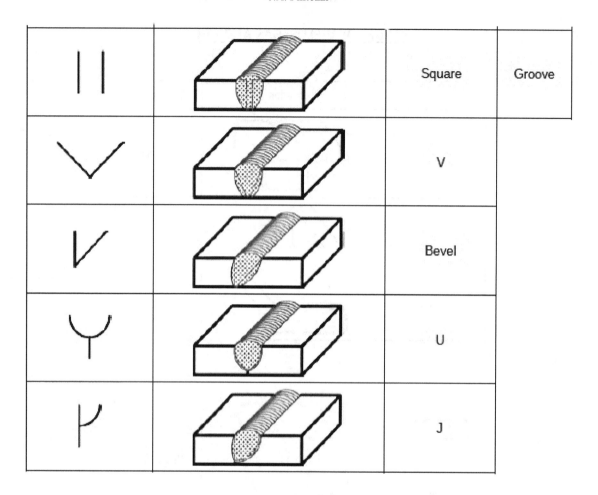

Groove welds are further divided among square, V, bevel, U, and J. Welds may be applied to one or both sides of a welded joint.

The Basic Weld Symbol: The most important part of the welding symbol is the basic weld symbol. One of the basic weld symbols indicating weld type is found on a line called the "reference line," attached with a leader whose arrow points to the joint where the weld is to be made.

Welding Symbols:

The most important part of the welding symbol are the basic weld symbols. One of the basic weld symbols indicating weld type is found on a line called the "reference line," attached with a leader whose arrow points to the joint where the weld is to be made. See the next image.

An example of the welding symbol is shown below. Welding symbols indicate the physical type of weld, such as a fillet or a seam weld, as well as other specifications related to the weld. Use of this symbol has been standardized by the American Welding Society in its publication *AWS Standard A2.4.* The symbol is shown in along with the locations of the various weld process indications for specifying type, size, contour, finish, and other characteristics.

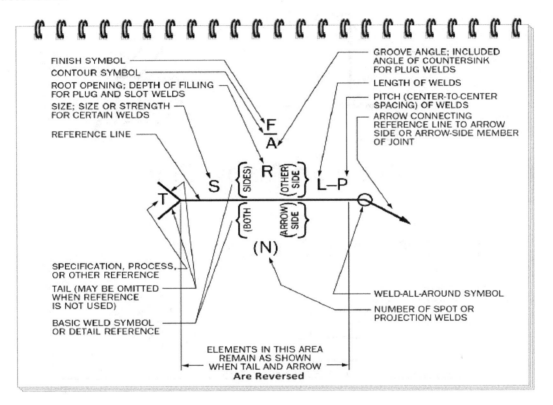

An explanation of the various specifications of the welding symbols follows. Note that the most common types of welds and uses of the welding symbols are detailed here. For a comprehensive explanation of all welding symbols used in industry, see the AWS standard.

The Basic Weld Symbol is the most important part of the welding symbol is the basic weld symbol. One of the basic weld symbols indicating weld type is found on a line called the "reference line," attached with a leader whose arrow points to the joint where the weld is to be made.

When a fillet basic weld symbol is used, the size of the weld leg is found to the left of the symbol, position "S" in the above image; all the positions noted below are shown in the following images. Fillet welds usually have equal-length legs, although legs of different length can be specified.

When a groove basic weld symbol is used, the size specification in position "S" indicates the depth of the groove (V, bevel, U, or J). Where no size is indicated, the groove is the full depth of the member.

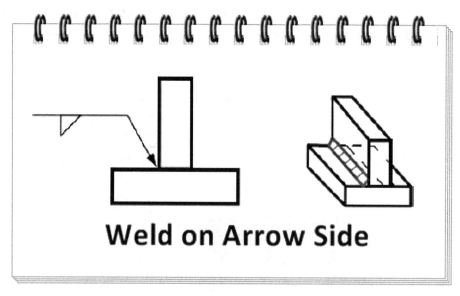

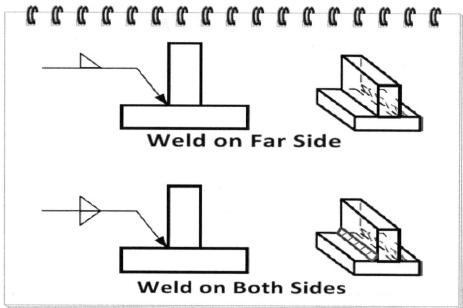

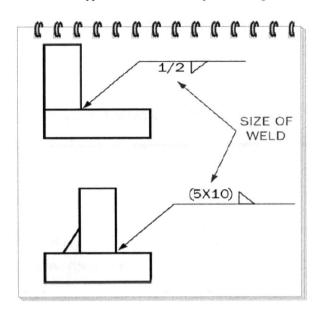

THIS WELDING SYMBOL INDICATES AN ALL AROUND WELD.

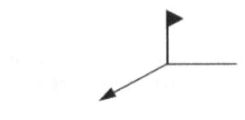

A FIELD WELD IS INDICATED HERE.

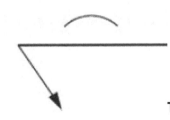

THIS IS THE WELDING SYMBOL FOR A CONVEX WELD.

THIS IS THE SYMBOL FOR A CONCAVE WELD.

Length and Pitch:

A pair of numbers separated by a dash (positions "L" and "P") indicates chain or intermittent welds, as shown below. The first number is the length of each weld; the second number is the pitch.

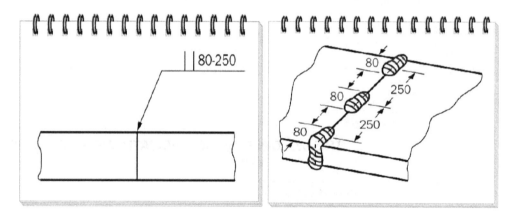

Finish Symbol:

The presence of a letter next to the contour symbol indicates the machining process that is to be used to achieve the indicated contour (C: chipping, G: grinding, M: machining, R: rolling, or H: hammering).

For Example: The symbol below indicates a V-Groove Weld, which requires machining.

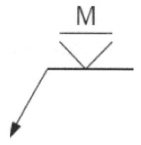

Many weldments are used without performing any secondary operations. However, precision parts may require additional operations. These operations are generally material-removal processes to produce accurate features and surfaces and improve surface finish on those areas requiring it.

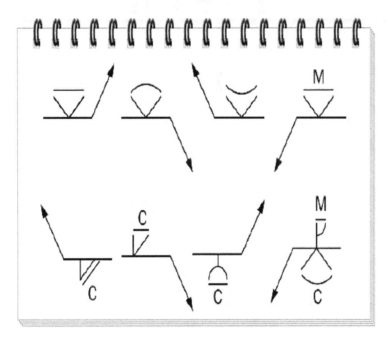

STANDARDS:

American industry has adopted a new standard, *Geometrical Dimensioning and Tolerancing,* ANSI Y14.5M-1982. This standard is used in all blueprint production whether the print is drawn by a human hand or by computer-aided drawing (CAD) equipment. It standardizes the production of prints from the simplest hand-made job on site to single or multiple-run items produced in a machine shop with computer-aided manufacturing (CAM) which we explained in chapter 2.

Industry standards for such things as threads and gears are referred to throughout the text. When a print refers to an item that has a standard, the entire standard applies. As an example, in a print that specifies a 1/4-20 UNC-2A thread, UN stands for "Unified," and it has an industry standard. The standard specifies all data pertaining to Unified threads, such as maximum and minimum diameters. These details therefore do not have to appear on a print. Industry standards are a very important part of manufacturing and keep prints uncluttered.

PART II

GD&T

J.A. Pintozzi

Chapter 1

GEOMETRIC DIMENSIONING and TOLERANCING

GD&T

Geometric dimensioning and tolerancing will be referred to as GD&T, throughout this book.

First, we will take a look at the history of GD&T, and the "Industrial Revolution." Drawings before the Industrial Revolution were elaborate pictures of parts that could not be mass produced, back then. Why is that you ask?

The Parts that were made pre-Industrial Revolution were unique and one of a kind. Each part was made as the craftsman saw the part. One machinist could view the drawing one way, while another machinist could view it another way. Mass production created a need for interchangeable parts, Fords assembly lines are perfect examples of the need for interchangeable parts. However, in order to manufacture parts that were so close to each other in size that they could be swapped, without making any modifications. Blueprints needed to be simplified and standardized, in a way that eliminated individual interpretations of the drawings and any guesswork on the part of the machinist. The industrial revolution made manufacturers realize that one of a kind parts, were not economical.

In 1966, we saw the first publication of the first GD&T standard, ANSIY14.5. Since then updates were published in 1982, 1994 and 2004.

While "Conventional Tolerancing" is suitable for some of our manufactured products, there are many others that require a higher degree of precision and tighter tolerancing. Geometric dimensioning and tolerancing provides us with a system for insuring a high degree of precision when machining. This, in turn guarantees that the machined parts, can easily be interchanged with each other. GD&T also allows the shop a high degree of repeatability when manufacturing precision parts, in production runs. There are standards and practices for creating technical drawings of mechanical parts and assemblies. The governing agencies responsible for setting the standards, includes the American Society of Mechanical Engineers (ASME). Then we have the International Standards Organization (ISO) and The American National Standards Institute (ANSI). There are a number of documents published by these organizations that cover various aspects of mechanical drawings. Remember a blueprint should communicate the following to the machinist:

- The Geometry of the part.

- Any Critical functional relationships.

- All Tolerances.

- Material, heat treat, surface coatings and special notes and instructions.

- Part documentation such as part number and drawing revision number.

When to use geometric tolerancing

- not needed if dimensional tolerances and the manufacturing process provide adequate control

- is needed
  - when part features are critical to function or interchangeability
  - when errors of shape & form must be held within tighter limits than normally expected from the manufacturing process
  - when functional gauging techniques are to be used
  - when datum references are required to ensure consistency between design, manufacture and verification operations
  - when computerization techniques in design and manufacture are used

Remember that Geometric Dimensioning and Tolerancing (GD&T) is an international language that is used on engineering drawings to accurately describe a part. It is a precise mathematical language that can be used to describe the size, form, orientation and location of part features.

 GD&T is also a design philosophy in which the part is defined based upon function in the final product. It works well with Concurrent Product and Process Development (CPPD) also called Simultaneous Engineering)

A design is created with inputs from marketing, customers, manufacturing, purchasing and any other area that has impact on the final production and use of the product. In the end, the ultimate goal of GD&T is to Provide the customer with what they want, when they want the product, how or the form they want the product in and at the price they want. Remember that the process of producing the product starts with the designer.

GD&T provides us with Improved Communications, Design uniformity, which allows design, production and inspection to all work from the same view. There are no arguments over what to do. This allows for better product design and the ability for designers to say what they mean instead of having engineers explain what to do.

GD&T provides us with a Increased Production Tolerance.

It gives us a Bonus Tolerance: which cuts costs during the manufacturing process. The Tolerance is based upon the parts functional requirements.

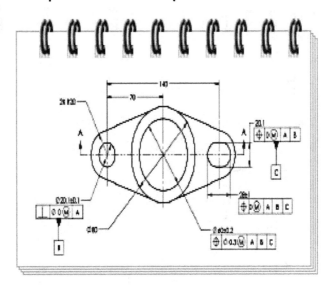

Basic GD&T Rules for Dimensions:

A dimension is a numerical value expressed in the appropriate units of measure and used to define the size, location, orientation, form or other geometric characteristics of a part.

- Each dimension should have a tolerance except those dimensions specifically identified as reference, maximum, minimum or stock size.

• Dimensions should be selected and arranged to suit the function and mating relationship of a part and not be subject to more than one interpretation.

GD&T is a combination of symbols and characters that supplements numeric dimensions and tolerances. Both numeric tolerances and geometric tolerances are used on the same drawing. GD&T provides a clear and concise standard for illustrating tolerances. The symbols are simple and easy to understand and have an exact meaning defined by industry standard. The system provides a precise method of specifying tolerances on industrial prints. Since GD&T is an accepted standard that is used throughout the manufacturing world, the symbols have the same meaning in any language. The purpose of this chapter is to provide a brief overview of the system as it pertains to basic print reading. Basically, GD&T allows us to dimension our drawings more accurately.

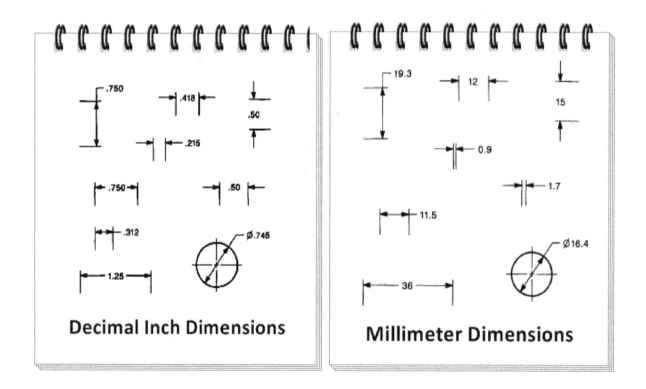

ANSI Y14.5M-1994. Fundamental Dimensioning Rules:

• Each dimension shall have a tolerance except those dimensions specifically identified as reference, maximum, minimum or stock size. These terms will be explained later on.

- **Dimensioning and tolerancing shall be complete so there is full definition of each part feature.**

- **Dimensions shall be selected and arranged to suit the function and mating relationship of a part and not be subject to more than one interpretation.**

- **These three rules establish dimensioning conventions.**

- **The drawing should define a part without specifying manufacturing methods.**

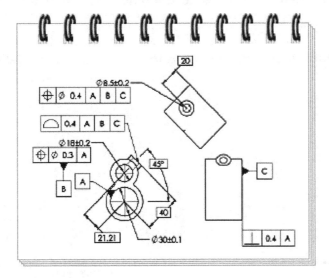

- **A 90 degree angle applies where centerlines and lines depicting features are shown on a drawing at right angles and no dimension is shown.**

- **A 90 degree basic angle applies where centerlines of features in a pattern or surfaces shown at right angles on a drawing, are located and defined by basic dimensions and no angle is specified.**

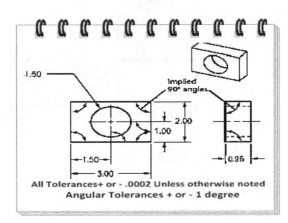

Note: These last two rules establish convention for implied 90 degree angles.

Unless otherwise specified, all dimensions are applicable at 20 degrees Centigrade (68 deg. F)

- All dimensions and tolerances apply in the free-state condition. This condition does not apply to non-rigid parts.

- Unless otherwise specified, all geometric tolerances apply to the full depth, length and width of the feature.

- These three rules establish default conditions for dimensions and tolerance zones.

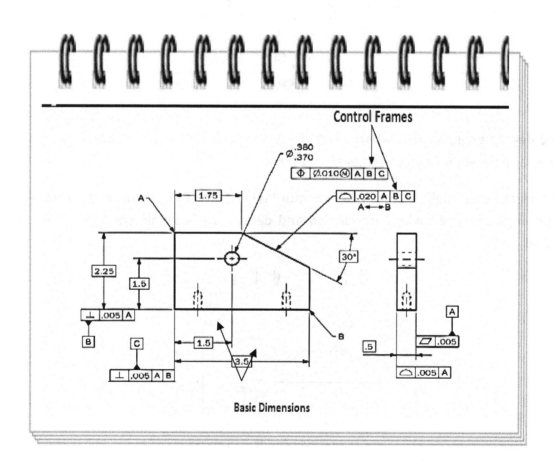

Basic Dimensions

Basic Dimensions:

The Dimensions in the image below are basic dimensions. Basic dimensions appear in conjunction with geometric tolerances. Basic dimensions have no tolerances, they describe the intended geometry of the part. Remember that "Basic Dimensions" on the print are not telling you that the part must machined perfectly; Instead, they should only alert you that the tolerance that describes the any allowable variation is found elsewhere on the print, such as in a feature control frame.

Basic dimensions describe the perfect size, shape, and location that are intended. They do not described any tolerances that are allowed. As you can see below, a basic dimension contains no tolerances.

$$\boxed{\varnothing\ .50}$$

Remember to study the print carefully to locate the tolerances for basic dimensions and other pertinent information, it is up to you to locate them. Just remember that basic dimensions are meant to describe the perfect size, shape, or location of a feature, not the allowable tolerances.

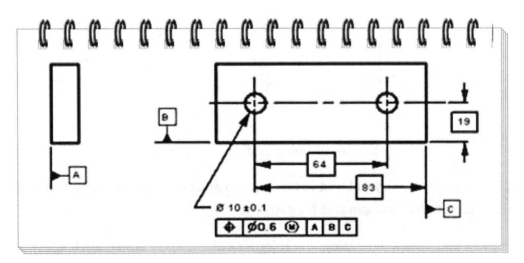

Basic Dimensions are used to define theoretically exact locations, orientation or true profile of part features or gage information. They define part features must be accompanied by a geometric tolerance. Basic Dimensions are normally enclosed in a rectangle or specified on the drawing as a general note. A basic dimension only locates a geometric tolerance zone. If a basic dimension describes a part feature, a geometric tolerance must be applied.

Basic dimensions may be used to specify datum targets. These dimensions are considered to be gage targets. That is the tolerance of the gage applies to this dimension.

Basic Dimensions define gage information and do not have a tolerance on the drawing. They are theoretically exact. But gage makers tolerance does apply.

Be sure you have the correct print for the part to be made or repaired. You want the print, which has not only the correct title, but also the correct assembly number. Never take a measurement with a rule directly from the print because the tracing from which the print was made may not have been copied from the original drawing perfectly and may contain scaling errors. Also, paper stretches and shrinks with changes in atmospheric conditions. Dimensions must be taken only from the figures shown on the dimension lines. Be very careful in handling all blueprints and working drawings. When they are not in use, place them on a shelf, in a cabinet, or in a drawer. Return them to the blueprint file as soon as the job is done. Blueprints and working drawings are always valuable and often irreplaceable. Make it a point never to mutilate, destroy, or lose a blueprint.

Basic dimensions may be used to specify datum targets. These dimensions are considered to be gage targets. That is the tolerance of the gage applies to this dimension.
Gage tolerance is an extremely small number. Many gages have tolerances in the thousandth, ten thousandth or greater amount.

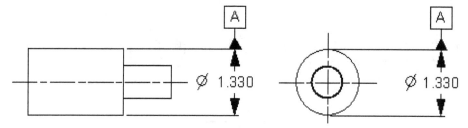

All dimensioning limits are absolute:

- That is all dimensions are considered to have a zero after the last true digit. (Ex. 1.25 means 1.250) This interpretation is important for gauging.

- For example a part with a dimension and tolerance of 1.0 +/- .1 would fail inspection at what Measure? (1.101 on the plus side and .899 on the low side)

ORIENTATION CONTROLS:

Orientation controls are used with features of size and surfaces. When they are applied to surfaces, they also control the form. However, they do not control the location of features. When applied to features of size, the center plane or axis of the feature is controlled, and the bonus tolerance is available. In the geometric system, there are three tools for controlling the orientation of features: parallelism, perpendicularity, and angularity.

Orientation controls provide full 3D control, but they can be made 2D by the note "LINE ELEMENTS" appearing beneath the feature control frame.

In the geometric system, there are three tools for controlling the orientation of features: parallelism, perpendicularity, and angularity.

| angularity | ∠ | perpendicularity | ⊥ | parallelism | // |

POSITION CONTROLS:

A position control is applied only to features having size, and it limits both location and orientation variation. Position can be used for both round features like holes and square or rectangular features like key-seats.

For example, when machining a round hole, the axis of the hole is controlled in all three dimensions. The hole will have a cylindrical-shaped tolerance zone extending the full depth of the hole. Note that the position tolerance requires the axis of the hole to be within the tolerance zone throughout the entire hole. This will affect both the location and tilt (orientation, or attitude) of the hole.

When machining a rectangular key-seat, the center plane of the key-seat width is controlled in all three dimensions with a tolerance zone. It's shape is the space that lies between two parallel planes.

This controls the location of the key-seat on its shaft axis, the tilt of the key-seat lengthwise, and the tilt of the key-seat depth wise.

FEATURES:

Parts are made up of features, which can be divided into two classes: features that have size and surfaces. Successfully interpreting prints with GD&T requires you to think about your parts in terms of their features, not only in terms of the dimensions appearing on the print.

Features that have size have a directly dimensioned size tolerance, which may be placed explicitly next to the dimension or implicitly on a title block or symbolic key. The maximum and minimum limits of size are given special names in the geometric system, which are helpful for understanding bonus tolerance.

The terms "MMC" and "LMC" are useful for calculating clearances between mating parts. More on MMC and LMC later.

A Feature is a general term applied to a physical portion of a part such as a surface, hole or slot. A feature may be considered a part surface. Parts are made up of features, which can be divided into two classes: features that have size and surfaces. Successfully interpreting prints with GD&T requires you to think about your parts in terms of their features, not only in terms of the dimensions appearing on the print. Features that have size have a directly dimensioned size tolerance, which may be placed explicitly next to the dimension or implicitly on a title block or symbolic key. The maximum and minimum limits of size are given special names in the geometric system, which are helpful for understanding bonus tolerance:

The terms "MMC" and "LMC" are useful for calculating clearances between mating parts. More on MMC and LMC later.

- Maximum material condition (MMC): The maximum size limit for an external feature or the minimum size limit for an internal future. In either case, whatever limit, if produced, results in the part having the maximum amount of material allowed.

- Least material condition (LMC): The minimum size limit for an external feature or the maximum size limit for an internal future. In either case, whatever limit, if produced, results in the part having the least amount of material allowed.

• Are comprised of part surfaces, or elements, that are internal part surfaces such as a hole diameter or the width of a slot.

• A feature of size dimension is a dimension that is associated with a feature of size. A non-feature of size dimension is a dimension not associated with a feature of size.

A fundamental rule of geometric dimensioning and tolerancing is that a size dimension controls not only the size of the feature, but also its form. Note: Any feature with size that is produced at its MMC is only acceptable only in its perfect form.

Table A-2

Symbol	Meaning
Ⓜ	Maximum Material Condition (MMC)
Ⓛ	Least Material Condition (LMC)
NONE	Regardless of Feature Size (RFS) Current Practice.
Ⓢ	Regardless of Feature Size (RFS) Old Style
▷	Translation
Ⓟ	Projected Tolerance Zone
Ⓕ	Free State
Ⓤ	Unequally Disposed Profile

Symbol	Meaning
Ⓘ	Interdependency
Ⓣ	Tangent Plane
⟨ST⟩	Statistical Tolerance
⟨CF⟩	Continuous Feature
CR	Controlled Radius
□	Square
[A]◄	Datum Feature Identifier

Table A-2

FEATURES OF SIZE:

Parts are made up of features, which can be divided into two classes: features that have size and surfaces. Successfully interpreting prints with GD&T requires you to think about your parts in terms of their features, not only in terms of the dimensions appearing on the print.

Features of size can have a directly dimensioned tolerance, the tolerance can be found directly next to the dimension or in the title block of the print. The maximum and minimum limits of size are given special names in the geometric system, which are helpful for understanding bonus tolerance.

Basically a feature of size has one cylindrical or spherical surface, or it can have a set of two opposed elements, or opposed parallel surfaces that are associated with a size dimension.

The following image is an example of features of size.

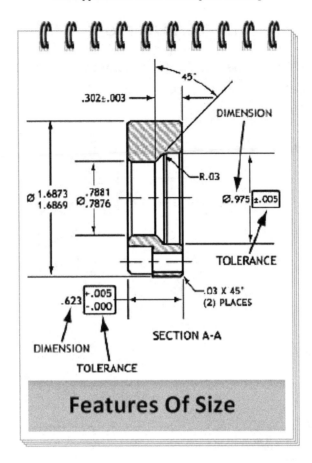

A feature is a general term applied to a physical portion of a part such as a surface, hole or slot. A feature may be considered a parts surface. Geometric tolerances can be displayed on a drawing by using a feature control frame.

INTERNAL FEATURES OF SIZE:

Internal features of size (FOS) are surfaces or elements that lie inside the part. An example would be the diameter of a hole or the width of a slot.

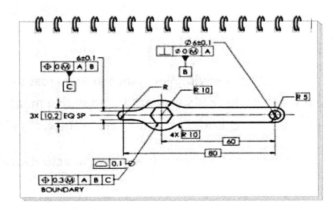

EXTERNAL FEATURES OF SIZE:

External features of size (FOS) are part surfaces or elements on the outside of the part. An example of an external feature of size could be the diameter of a shaft or the overall width or height of a planar part.

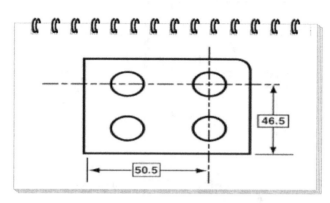

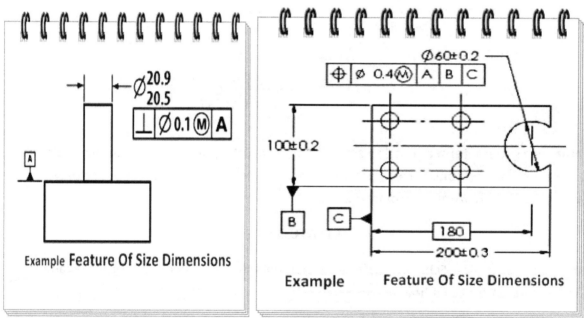

Example Feature Of Size Dimensions

Example Feature Of Size Dimensions

ACTUAL LOCAL SIZE:

An actual local size is the value of any individual distance at any cross section of a feature of size. The actual local size is a two point measurement taken with a measurement instrument, such as a caliper or micrometer. It is checked at a point along the cross section of the part.

Note: A feature of size can have several different values for an actual local size.

1. The geometric characteristic symbol: This symbol indicates what type of geometric tolerance applies; see table A-1 below . In this example, the first box tells that we are dealing with perpendicularity.

2. Next, you have the shape of the tolerance zone, (Optional): The shape will be identified by one of the following: a square symbol, a diameter symbol, or a spherical diameter (rarely), or it can be left blank. When no symbol is given, the shape of the tolerance zone is inferred from the shape of the feature.

3. The total size of the tolerance zone: This is always given in the same linear dimension units used on the rest of the print—inches or millimeters. In the image above, we can see the size of our tolerance zone, .008.

4. The material modifier symbol (optional): The symbol will be one of these three: the MMC symbol, the LMC symbol, or the S symbol or no symbol at all. For MMC, bonus tolerance is available on an MMC basis; and for LMC, bonus tolerance is available on an LMC basis. If no symbol is given or if the S symbol is used, there is no bonus tolerance allowed.

A fundamental rule of geometric dimensioning and tolerancing is that a size dimension controls not only the size of the feature, but also its form. Any feature with size that is produced at its MMC is only acceptable if its form is perfect.

Material Condition and Material Condition Modifiers:

GD&T allows for certain modifiers to be used in specifying tolerances at various part feature conditions. These conditions may be the largest size of the feature, the smallest size of the feature or actual size of that feature. Material conditions may only be used when referring to a feature of size.

Material Condition Modifiers are used to refer to a feature in its largest or smallest condition, or to refer to it regardless of feature size. There are three Material Condition Modifiers:

– Maximum Material Condition (MMC): MMC might be used, for example, to refer to the largest pin or the smallest hole.

– Least Material Condition (LMC): LMC could refer to the smallest pin or the largest hole.

– Regardless of Feature Size (RFS): RFS could refer to any increment of feature size of any feature within its size tolerance.

- Features of size, which include datum features, have size tolerances.

- The size condition or material (the amount of metal) condition, can vary from maximum material condition (MMC) to the least material condition (LMC).

- If the center planes or axes of a feature of size are controlled by geometric tolerances, a modifying symbol is placed in the feature control frame. Either MMC or LMC can be applied to the tolerance value.

- If a symbol is not present in the control frame, the regardless of feature size is implied (RFS).

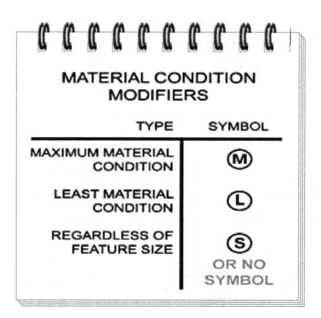

Maximum material condition is when a feature of size contains the maximum amount of material everywhere within the stated limits of size. This could be the largest shaft diameter or smallest hole.

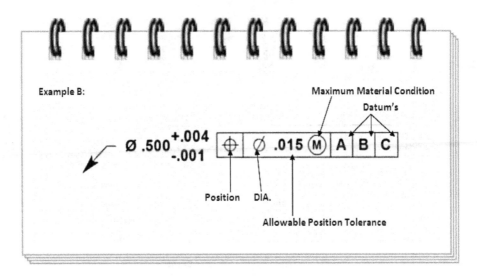

(Example B)

• If a hole, for instance, has the following size and geometric control, and the hole measures .502.

• It would be incorrect to use a bonus tolerance of .003 (.502 - .499(MMC)) if the hole is not perfectly oriented to the Datum's.

• If the hole is out of perpendicular to datum A by .002, for instance, the bonus that may be used is reduced by that amount. The bonus would be merely .001 and the allowable position tolerance = .016

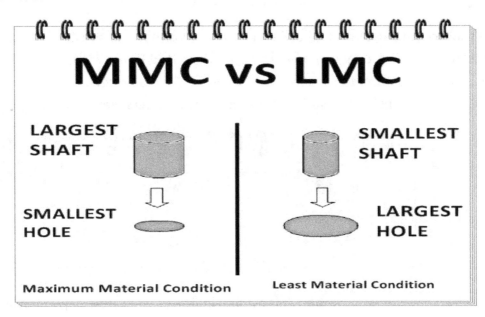

Geometric Characteristic Symbols				Modifying Symbols	
Type of Tolerance		Characteristic	Symbol	Term	Symbol
For individual features	Form	Straightness	—	At maximum material condition	Ⓜ
		Flatness	▱	At least material condition	Ⓛ
		Circularity (roundness)	○	Projected tolerance zone	Ⓟ
		Cylindricity	⌭	Free state	Ⓕ
For individual or related features	Profile	Profile of a line	⌒	Tangent plane	Ⓣ
		Profile of a surface	⌓	Diameter	⌀
For related features	Orientation	Angularity	∠	Spherical diameter	S⌀
		Perpendicularity	⊥	Radius	R
		Parallelism	∥	Spherical radius	SR
	Location	Position	⌖	Controlled radius	CR
		Concentricity	◎	Reference	( )
		Symmetry	⌯	Arc length	⌒
	Runout	Circular runout	↗	Statistical tolerance	⟨ST⟩
		Total runout	⌰	Between	↔

THE CONTROL FRAME:

Remember that the rectangular box is called a feature control frame and indicates a geometric tolerance. The proper way to read a feature control frame, is from left to right. As you can see, a feature control frame consists of a series of symbols, letters and numbers. A feature control frame consists of a series of symbols and numbers. From left to right:

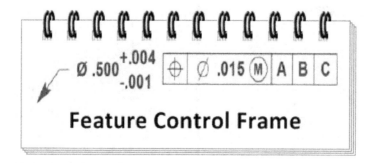

Feature Control Frame

1. The geometric characteristic symbol: This symbol indicates what type of geometric

tolerance applies. In the image above the position symbol is used.

2. **The shape of the tolerance zone (optional):** The shape will be identified by one of the following: the square symbol, diameter symbol, spherical diameter (rarely), or blank. When no symbol is given, the shape of the tolerance zone is inferred from the shape of the feature. The image above tells us that the tolerance zone is a diameter.

3. **The total size of the tolerance zone:** This is always given in the same linear dimension units used on the rest of the print—inches or millimeters. Looking at the image we see that we have a .015 tolerance zone.

4. **The material modifier symbol (optional):** The symbol will be one of these three: the MMC symbol, the LMC symbol, or the RFS symbol (prior practice). For MMC, bonus tolerance is available on an MMC basis; and for LMC, bonus tolerance is available on an LMC basis. If no symbol is given or if the RFS symbol is used, there is no bonus tolerance allowed.

The Ⓜ symbol indicates a Maximum Material condition (MMC).

5. **The datum reference frame (mandatory for some geometric characteristics, excluded for some, optional for some):** This will be one or more datum references listed in order of precedence (primary, secondary, etc.). Finally the above control frame references datum's A, B and C.

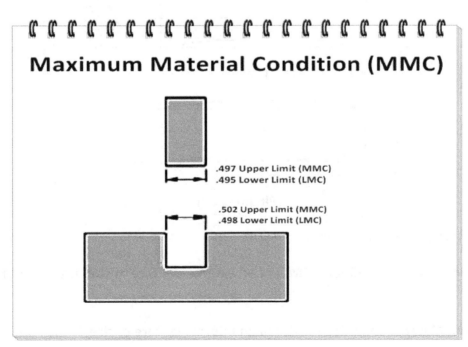

The size of the pin in the following two images is allowed to vary from 15.98 to 16. Variation from perfect form (16.) is only allowed to the extent that the size varies from MMC. Some form variation is allowed (bow, necking, barreling, out-of-round, etc.), but only variation that does not push the material of the pin beyond its 16-millimeter size.

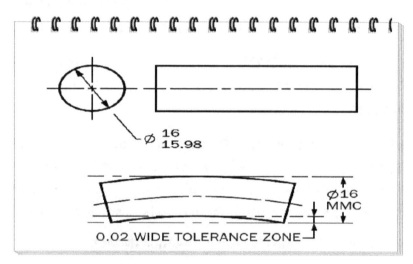

(Pin Image 1)

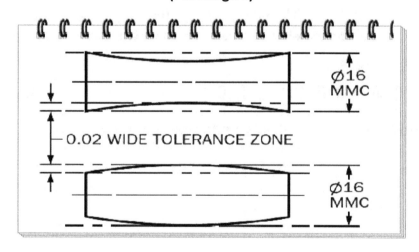

(Pin Image 2)

Some important points to remember:

The Maximum Material Condition (MMC) of an external feature of size, is the features largest size limit.

The Maximum Material Condition (MMC) of an internal feature of size, is the features smallest size limit.

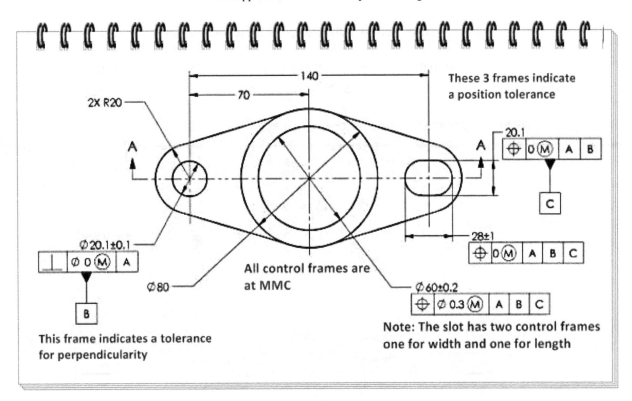

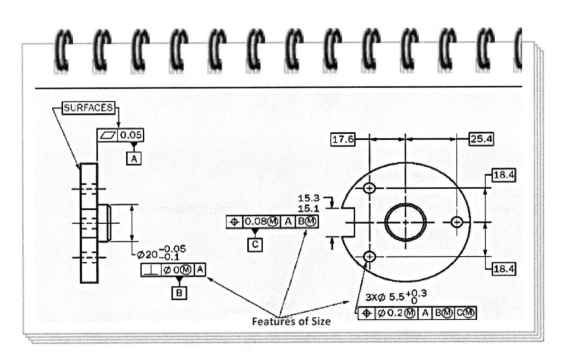

(Examples of Features of size)

LEAST MATERIAL CONDITION (LMC):

The least material condition (LMC), is the condition in which a feature of size contains the least amount of material everywhere within the stated limits of size.

LMC would be the smallest shaft diameter or the largest hole diameter.

A few key points to remember are, first that the least material condition (LMC) for an external feature of size, is the smallest size limit. And secondly the least material condition (LMC) for an internal feature of size, is its largest size limit.

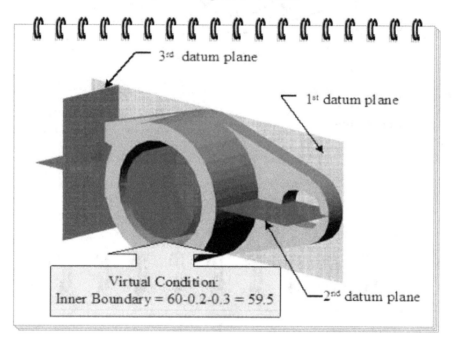

REGARDLESS OF FEATURE SIZE (RFS):

• Regardless of feature size (No Symbol) indicates that a geometric tolerance applies at any increment of size, of the feature within its size tolerance.

• RFS uses no symbol.

• RFS is the default condition for all geometric tolerances.

• RFS is always used for runout, concentricty and symmetry controls.

• RFS is used to control wall thickness variation between external and internal features.

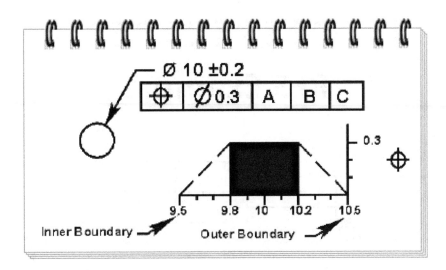

In the above illustration the regardless of feature size is implied. Note: the RFS is the most expensive control to use. Even though the size is not less than 9.8 or greater than 10.2 mm., the hole might act like 9.5 mm. to 10.5 mm.

Holes made from 9.5mm. to 9.8mm. and from 10.2mm. to 10.5mm. should be rejected even though they may still function.

RFS RULE 2:

RFS applies to the individual tolerance, datum reference or both where no modifying symbol is used. MMC or LMC must be specified on the drawing where it is required.

Remember that GD&T modifiers provide us with additional information about the part. This information helps us to understand the meaning of the drawing and how to machine the part. Now that we have discussed MMC, LMC and RFS, we will take a look at a few more.

P: Projected Tolerance Zone:

A projected tolerance zone is a tolerance zone that extends beyond a feature by a specified distance. Projected tolerance zones help ensure that mating parts fit during assembly.

T: TANGENT PLANE:

A tangent plane notes that only the tangent plane of the toleranced surface need be within the tolerance zone.

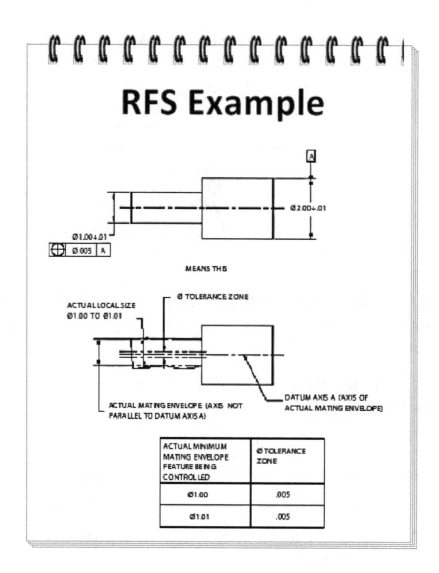

MODIFIERS:

⌀ The Diameter Symbol when used in the feature control frame, denotes the shape of the tolerance zone. When used outside of the feature control frame it replaces the word Diameter.

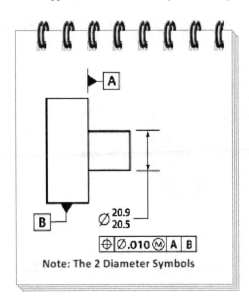

Note: The 2 Diameter Symbols

R = Radius and CR = Controlled Radius, they are used outside of the feature control frame.

RADIUS:

A radius is a straight line extending from the center of an arc or circle to its surface. The symbol for a radius is R. The radius creates a tolerance zone defined by two arcs. The actual surface of the part must lie within the tolerance zone, created by the two arcs. The two arcs are the minimum and maximum radii.

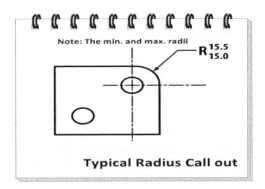

Typical Radius Call out

CONTROLLED RADIUS:

A Controlled Radius (CR) is a radius with no flats or reversals. The controlled radius is drawn the same way as Radius (R), the only difference is that a CR cannot have any flats or reversals within the minimum and maximum radii. The radius is marked as CR on the drawing just as R is used for a standard radius.

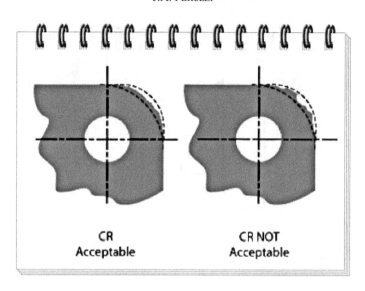

Limits of Accuracy:

All machinists must work within the limits of accuracy, as specified on the blueprint. As a machinist, understanding tolerance and allowance will help you in making good parts. These terms, "Tolerance" and "Accuracy" may seem to be related, but each has a very different meaning and application. In the following pages, you will learn the meanings of these terms and the difference between them.

CHAPTER 7

TOLERENCE:

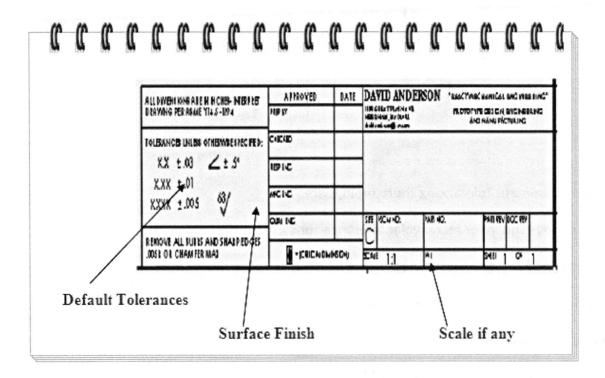

- Remember that "Coordinate Tolerancing" is a dimensioning system where a part feature is located by a means of rectangular dimensions with given tolerances. Coordinate tolerancing does have its problems, it lacks the completeness required to manufacture parts efficiently and economically.

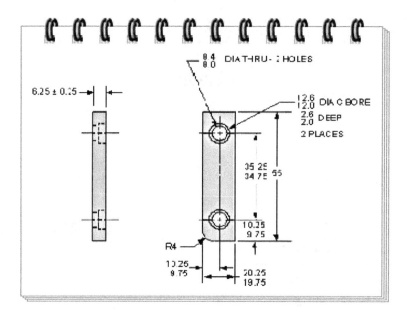

Some Coordinate Tolerancing shortcomings are:

• It employs Square or rectangular tolerance zones.

• It has fixed size tolerance zones.

• Its instructions for inspection are ambiguous.

• It doesn't aid in locating Part Features.

• It doesn't control angular relationships.

• It doesn't define the form of part features.

One Advantage of GD&T over Coordinate tolerancing is an Increased Tolerance Zone. The tolerance is no longer square but fits the geometry of the feature.

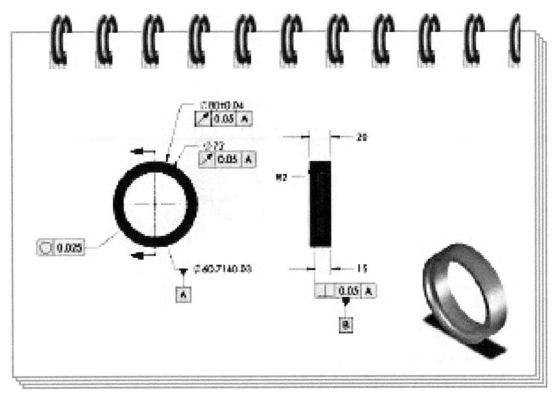

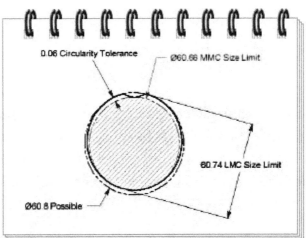

A Fixed Size and Square Tolerance Zones

• Circle is 0.71 diameter

• Square is 0.5 each side

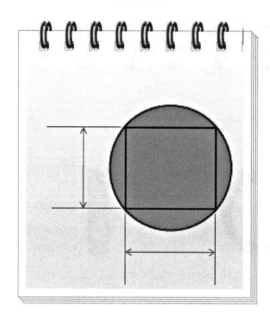

GD&T provides Ambiguous Instructions for Inspection

Where Coordinate Tolerancing allowed the inspector to "inspect" the part from whatever perspective the inspector deemed appropriate.

GD&T gives a defined process for inspection. GD&T uses the datum system to eliminate inspection confusion.

The parts described on two-dimensional manufacturing prints are in fact three-dimensional objects. Features on three-dimensional objects have up to four different types of variation from their "ideal" design: Size • Location • Orientation • Form Geometric tolerances Versus Size Tolerances. GD&T does not replace conventional forms of tolerances; instead, it supplements them.

Conventional, or "plus-or-minus," tolerances are effective at describing limits on allowable variation for sizes, but when they are applied to locate features or to orient features, there is often ambiguity. GD&T eliminates this ambiguity, the shapes of surfaces are not controlled by conventional tolerancing.

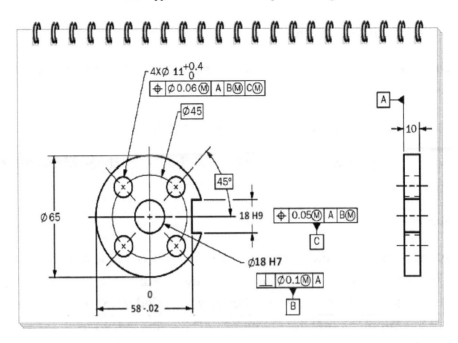

The above print is defined with plus-or-minus (±) dimensions. Using the ± dimensions for the size of the hole and the thickness of the plate is a clear and acceptable method for defining these features.

- Note that the remaining ± dimensions on the part are not clear and are subject to misinterpretations.

Remember that tolerance is the total amount that a specific dimension is allowed to vary from the specified dimension. The tolerance is the difference between the maximum and minimum limits of the dimension. Working to an exact specific dimension is impractical and all but impossible in most cases. So the designer will add to the specific dimensions, an allowable variation in size, the Tolerance. The amount of variation or limit of error permissible is indicated on the drawing as plus or minus a given amount. such as + 0.003 or - 0.003. The difference between the allowable minimum and the allowable maximum dimension is your tolerance.

- Large variations can affect the functionality of the part.

- A small variation will affect the cost of the part. Holding tighter tolerances will require precise machining and the inspection and rejection of parts.

Tolerances become important:

- When the part is part of an assembly, many times parts will not fit together, if their dimensions do not fall within a certain range of values.

• When interchangeability is required, such as a replacement part. It must be a duplicate of the original part within certain limits of deviation.

• Parts with tight tolerances will often require special methods of machining.

Note that the relationship between functionality and the size or shape of an object will vary from part to part.

As a rule of thumb:

• Implicit Tolerances (General Tolerances) are specified when all the dimensions in the drawing are using the same tolerance. Tolerances may be expressed Implicitly, as a general tolerance applied to all dimensions unless otherwise specified. Implicit tolerances are given in the title block or in a supplemental document or CAD file referenced from the print. Implicit tolerance means that no tolerance is shown with the dimension because it is stated either elsewhere on the print or document. Dimensions that are not basic have implicit tolerances. However, some prints may include a note indicating that all untoleranced dimensions are basic.

```
1 EXCEPT WHERE STATED OTHERWISE
  TOLERANCES ON DIMENSIONS ±.010


2 UNLESS OTHERWISE SPECIFIED
  ±.007 TOLERANCE ON MACHINED DIMENSIONS
  ±.10 TOLERANCE ON CAST DIMENSIONS
  ANGULAR TOLERANCE ±.1°
```

Notes like the one above are sometimes used to reduce the number of dimensions on a drawing and to increase drawing clarity.

Almost all industrial prints have a title block that contains a set of tolerances, these tolerances are called "general tolerances," "title block tolerances," or just implicit tolerances. They apply to all dimensions that do not have a direct tolerance or a geometric tolerance and are not covered by a separate note.

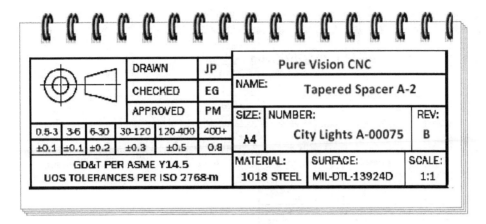

Above is an image of a metric title block with ISO 2768-m tolerance used Some angle tolerances are implied when their dimensions are also implied and do not appear on the print: An angle of 0, 90, or 180 degrees is implied where center lines or surfaces are shown on orthographic views as parallel or at right angles to one another. The tolerance on the implied angle is found on the title block.

NOTE: A basic angle of 0, 90, or 180 degrees is implied where center lines or surfaces are shown on orthographic views as parallel or at right angles to one another and geometric tolerances are applied.

GENERAL TOLERANCES:

Permissible Tolerances for Linear Dimensions, mm

	0.5–3	3–6	6–30	30–120	120–400	400–1000	1000–2000
Fine	±0.05	±0.05	±0.1	±0.15	±0.2	±0.3	±0.5
Medium	±0.1	±0.1	±0.2	±0.3	±0.5	±0.8	±1.2
Coarse	±0.2	±0.3	±0.5	±0.8	±1.2	±2	±3
Very coarse		±0.5	±1	±1.5	±2.5	±4	±6

Permissible Tolerances for External Radii and Chamfer Dimensions, mm

	0.5–3	3–6	Over 6
Fine	±0.2	±0.5	±1
Medium	±0.2	±0.5	±1
Coarse	±0.4	±1	±2
Very coarse	±0.4	±1	±2

Permissible Tolerances for Ranges of Shorter Side Lengths of the Angle, mm

	Up to 10	10–50	50–120	120–400	Over 400
Fine	±1°	±0°30'	±0°20'	±0°10'	±0°5'
Medium	±1°	±0°30'	±0°20'	±0°10'	±0°5'
Coarse	±1°30'	±1°	±0°30'	±0°15'	±0°10'
Very coarse 1+	±3°	±2°	+1 ±1°	±0°30'	±0°20'

Explicit Tolerances:

- Explicitly as direct limits or tolerance values applied directly to a dimension. As a geometric tolerance that may be associated with one or more basic dimensions.

An explicit tolerance means that a tolerance value is shown with the dimension. There are several ways in which explicit tolerances are expressed. No matter how the tolerance is displayed on the print, the maximum allowable value can be calculated as the dimension plus the plus tolerance, and the minimum allowable value can be calculated as the dimension minus the minus tolerance.

The dimension is given first and followed by the "plus-or-minus" (±) symbol and then the tolerance value, this is known as an Equal Bilateral tolerance.

Unequal bilateral tolerances Unilateral Unequal bilateral tolerances

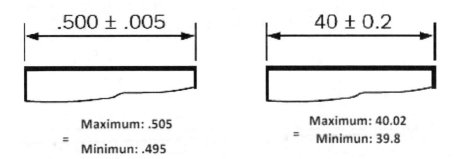

Unequal bilateral tolerances Unilateral:

The dimension is given first, followed by the tolerance displayed on two rows. Unilateral tolerancing is similar to unequal bilateral tolerancing, but with either the plus tolerance or the minus tolerance value equal to zero.

Unilateral tolerancing with a positive tolerance is referred to as "oversized"; unilateral tolerancing with a negative tolerance is referred to as "undersized."

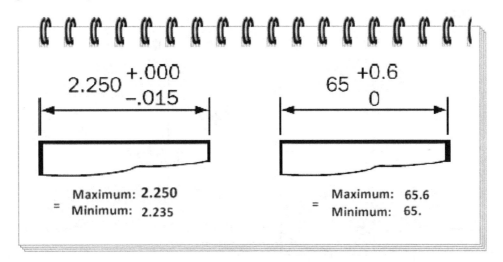

Unilateral tolerances Unequally Disposed:

The dimension is given first, followed by the tolerance displayed on two rows. Unequally disposed tolerancing is similar to unequal bilateral tolerancing and unilateral tolerancing, but with two positive tolerance values or two negative tolerance values.

Unequally disposed tolerancing with positive tolerance values is referred to as "oversized"; unequally disposed tolerancing with negative tolerance values is referred to as "undersized."

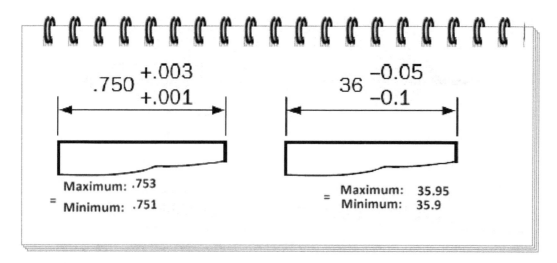

Tolerances may be shown as maximum and minimum dimensions and are called limits. Limits display a permissible range in the form of direct dimensions that do not require additional calculations. Limits are generally expressed with the larger dimension above the smaller dimension; however, limits for internal dimensions may have the smaller dimension above the larger dimension.

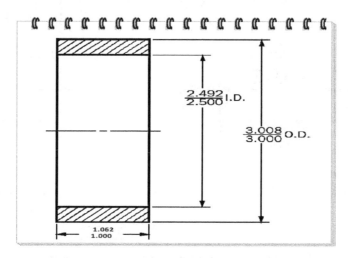

A MIN or MAX after a dimension indicates the minimum or maximum value permitted, with no limitation on the other limit. Single limits are used where the unspecified limit can be zero or approach infinity and the feature will still satisfy the design intent. The radius in the figure below cannot be larger than .125. It could be produced smaller, even zero for a sharp corner.

In the figure below the .500 Dia. hole must be at least .750 deep. The hole can be deeper, even be produced as a through hole.

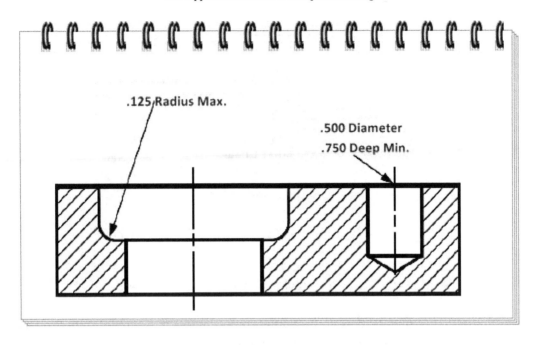

NOTE the following tolerance examples:

- **Tolerances for Inch Linear Dimensions:**

$.750 \pm .005$ 24 ± 0.1 $45°30' \pm 0°30'$
Inch Linear   Millimeter Linear   Angular

- **Equal Bi-lateral Tolerances:**

$.750^{+.002}_{-.003}$ $24^{+0.08}_{-0.20}$ $45.5°^{+0.2°}_{-0.5°}$
Inch Linear   Millimeter Linear   Angular

- **Un-equal Bi-lateral Tolerances:**

$.750^{+.000}_{-.004}$ $24^{0}_{-0.2}$ $45.5°^{0}_{-0.5°}$
$.750^{+.004}_{-.000}$ $24^{+0.2}_{0}$ $45.5°^{+0.5°}_{0}$
Inch Linear   Millimeter Linear   Angular

• **Unilateral Tolerances:**

$$.750{}^{+.003}_{+.002} \qquad 24{}^{+0.35}_{+0.20} \qquad 45.5°{}^{+1.5°}_{+0.5°}$$

$$.750{}^{-.001}_{-.010} \qquad 24{}^{-0.15}_{-0.25} \qquad 45.5°{}^{-0.5°}_{-1.0°}$$

      Inch Linear   Millimeter Linear   Angular

Unequally Disposed Tolerances:

$$\begin{array}{ccc} .752 & 24.08 & 46°30' \\ .747 & 23.8 & 45°30' \end{array}$$

      Inch Linear   Millimeter Linear   Angular

• On bilateral, unequally disposed, and limit tolerances, both the plus and minus values have the same number of decimal places as the dimension.

• On unilateral tolerances, either the plus or minus value is zero; a plus or minus sign is used followed by the decimal point, and the same number of zeroes is shown in the tolerance as digits in the dimension.

• On basic dimensions, trailing zeroes are not used.

Millimeter Linear Dimensions:

• Tolerances for millimeter linear dimensions are given in millimeters. On bilateral, unequally disposed, and limit tolerances, both the plus and minus values have the same number of decimal places, with trailing zeroes added if necessary.

• On unilateral tolerances, either the plus or minus value is zero; a single zero is shown without a plus or minus sign.

• On basic dimensions, trailing zeroes are not used.

Angular Dimensions:

• Tolerances for angular dimensions are given either in decimal degrees or in degrees, minutes, and seconds to match the units of the dimension. The respective units are given for both the dimension and the tolerance.

• For angle dimensions, both the plus and minus values have the same number of decimal places, with trailing zeroes added if necessary.

Angular Surfaces:

Angular surfaces may be indicated by: Two linear dimensions and tolerances Three linear dimensions and tolerances One linear dimension and tolerance, plus one angular dimension and tolerance Angular surfaces indicated on prints with a directly toleranced linear dimension and angular dimension, as seen below.

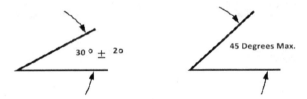

must be produced within a tolerance zone represented by two nonparallel planes.

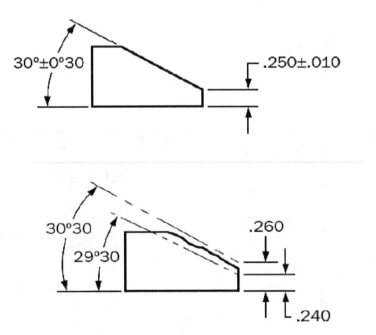

With tapers and slopes, the tolerance is applied to the first number of the ratio.

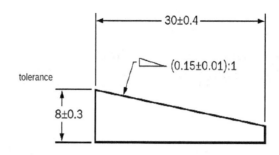

Tolerancing an angular surface with a linear and angular tolerances.

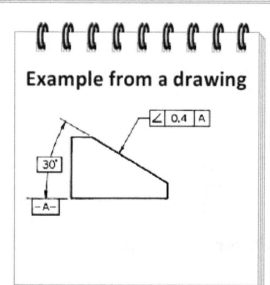

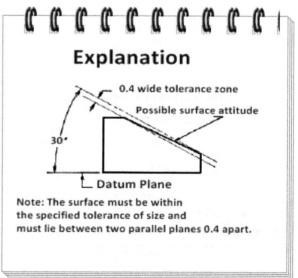

Conical Taper tolerances:

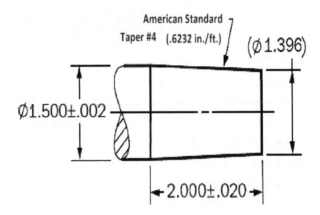

Conical tapers include standard machine tapers as described by ASME B5.10 and may be indicated by taper name and number; see the above image. Taper name is inches per foot.

Conical tapers may also by indicated by: Two diameter dimensions and tolerances, plus a linear dimension and tolerance.

One diameter and tolerance, a taper dimension and tolerance, and a linear dimension and tolerance.

Radius Tolerances:

A directly toleranced radius dimension indicates that the part surface must lie between two arcs drawn at the maximum and minimum radii; see the image below.

A controlled radius symbol, "CR," indicates that the part surface must not have any reversals in its contour.

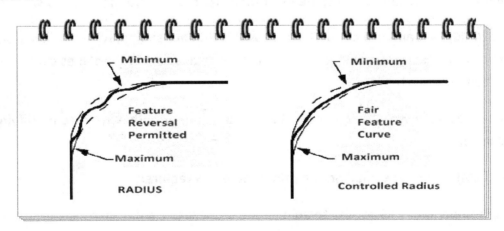

NOTE: When tolerances are not actually specified on a drawing, fairly concrete assumptions can be made concerning the accuracy expected. by using the following principles. For dimensions which end in a fraction of an inch. such as 1/8, 1/16, 1/32, 1/64. consider the expected accuracy to be to the nearest 1/64 inch.

When the dimension is given in decimal form, the following applies: If a dimension is given as 2.000 inches, the accuracy expected is +0.005 inch: or if the dimension is given as 2.00 inches, the accuracy expected is +0,010 inch. The +0.005 is called in shop terms, "plus or minus five thousandths of an inch." The + 0.010 is called "plus or minus ten thousandths of an inch.

Tolerances are specified by:

• Size: Limits specifying the allowed variation in each dimension; Length, width, height, diameter etc. are given on the drawing.

• Geometric Tolerancing: which allows for the geometry of a part separate from its size.

• GD&T uses special symbols to control the different geometric features of a part.

Bonus Tolerance:

In some cases, the geometric tolerance given in a feature control frame can actually be made larger during inspection, allowing more variation than the print would seem to indicate. This tolerance is sometimes referred to as "bonus tolerance," because it is added to the geometric tolerance in the feature control frame to get the total tolerance used to verify the feature. Bonus tolerance maybe indicated on the print for a size feature, but never for a surface.

The use of MMC, LMC and RFS us limited to features that are subject to variations in size. They can be datum features or other features that have axes or center planes controlled by geometric tolerances.

• RFS applies to an individual tolerance, a datum reference or both, where no modifying symbol is used.

• MMC and LMC will be specified on the print where it is required.

• For position tolerance RFS applies to an individual tolerance, a datum reference or both.

• When a geometric tolerance is applied as RFS, the geometric tolerance is independent of the actual size of the feature. The tolerance is limited to the specified value regardless of the features actual size. So referencing a datum feature with RFS means that it is necessary to center its axis or center plane, regardless of the features actual size.

The Following are the Characteristics Symbols That Are Used with GD&T.

Form Controls	—	Straightness
	▱	Flatness
	○	Circularity
	⌭	Cylindricity
Orientation Controls	∥	Parallelism
	⊥	Perpendicularity
	∠	Angularity

Location Controls	⊕	Position
	◎	Concentricity
	≡	Symmetry
Profile Controls	⌒	Profile of a line
	⌒	Profile of a surface
Runout Controls	↗	Circular runout
	↗↗	Total runout

MODIFYING SYMBOLS:

As a reminder; the first Feature control frame box (FCF) is a symbol, the second contains the tolerance value, and the third through fifth usually contain the Datum callouts. The second box of a Feature Control Frame may also contain a modifying symbol directly after the tolerance value. These are special symbols that can exist in any FCF box except the first. The first FCF box is reserved for *control symbols* already listed. Some of the *modifying* symbols are listed here:

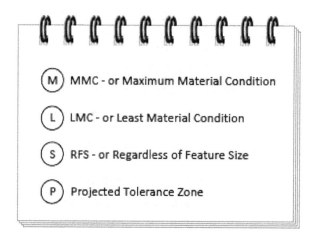

MMC and Projected Tolerance Zone will be covered briefly because they are two of the more useful modifying symbols. MMC is valuable for allowing more part variance during fabrication while ensuring that the parts will always assemble properly. Projected Tolerance Zone is used to control the location of a hole *beyond* the parts' physical edges.

MAX MATERIAL CONDITION (MMC) DEFINITION, PRACTICE:

Manufacturing tolerances are used when individual parts are made. When multiple parts must fit together to make an assembly, these individual part tolerances must be properly controlled. If they are not, then the part may not fit into the assembly and will need to be reworked or discarded. In order to maximize the allowable manufacturing variability while ensuring proper fit up, MMC may be used.

Maximum Material Condition, or MMC, is as it sounds - the most material. Consider a bolt going through a hole. The MMC bolt is the largest diameter it can be within its tolerance.

However, a MMC hole is the smallest hole, which also represents the most material. By including the MMC modifying symbol in a feature control frame, the size of the hole (diameter) now comes into play when determining *positional* tolerance. Let's take a look at an example.

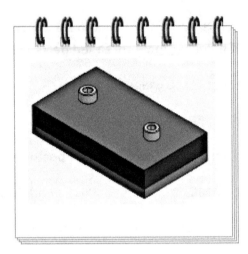

Here we have a two plate assembly; How do we allow more manufacturing tolerance in the TOP PLATE?

Notice that these parts will assemble even when the holes are at MMC, or the smallest hole. If both of the holes in the TOP RED PLATE were made at the large end of the tolerance, Ø.29, there would then be extra clearance. This is the principle behind allowing extra manufacturing tolerance with MMC, by making the hole larger its position can drift more.

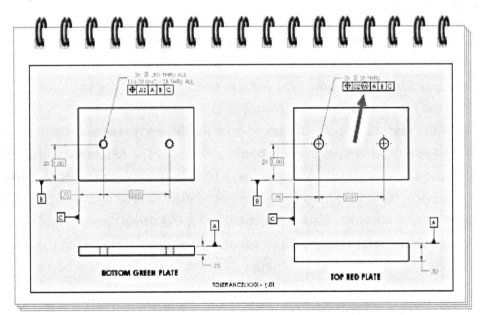

The drawing shown above represents the GD&T version, with MMC for the two holes in the TOP PLATE. Similar tolerances are used for the hole's position and diameter. The difference being the circled M in the TOP PLATE GD&T Feature Control Frame. This circled M means MMC applies to these two holes.

Since a threaded hole will follow its own strict size and tolerance definitions, MMC is not to be used on threaded or tapped holes, or male threaded parts.

Follow these steps to verify proper MMC usage:

1. Determine all of the holes' diameters at MMC.

2. With all of the holes at MMC (smallest), no *additional* positional tolerance is permitted.

3. For every .001" a hole is over MMC, it is allowed .001" of additional position tolerance.

4. Step #3 is valid up to LMC (Least Material Condition), or the largest hole within tolerance.

With MMC on a drawing the machinist has a choice. Make the holes smaller in diameter (within tolerance) and get little-to-no *extra* positional tolerance, or make the holes larger in diameter and get additional position tolerance. If MMC is applied and followed properly then the mating parts will always assemble even though extra positional tolerance may have been allowed during fabrication.

This example uses only two holes, but these same MMC principles apply to any bolt pattern be it square, rectangular, or round.

Remember A bonus tolerance is indicated in one of two ways, on a maximum material condition (MMC) basis or on a least material condition (LMC) basis. The basis for the bonus tolerance is indicated by the symbol in the feature control frame after the geometric tolerance. With the MMC modifier, the geometric tolerance given on the drawing applies when the size of the feature actually produced is its maximum material condition; for a feature produced at any other size, there is bonus tolerance. The RFS symbol after the geometric tolerance, or no symbol, indicates there is no bonus tolerance for the feature. One cylindrical or spherical surface or a set of two opposed elements or opposed parallel surfaces associated with a size dimension. Geometric tolerancing is a three-dimensional (3D) language. The geometric tolerance creates a zone of tolerance within which the feature must lie. Tolerance zones come in a variety of shapes depending on the control applied and the feature.

Tolerance zones can be three-dimensional or two-dimensional. Two dimensional (2D) tolerance zones are used when geometric control is more important in one direction than another. They apply to a cross-sectional element of a feature, such as straightness applied to a pin's surface.

Straightness is measured in one direction only, lengthwise; not around the pin, which is never intended to be straight.

The rectangular box is called a feature control frame and indicates a geometric tolerance. The feature control frame is read from left to right.

The calculation involves only two numbers: the actual produced size of the feature and either its MMC or LMC. The MMC or LMC can be calculated from the dimension and tolerance on the print; the actual produced size of the feature is measured at inspection. Only one calculation is used, depending on whether the feature is internal or external and on what modifier is used; see table below.

Calculating Bonus Tolerance and Total Tolerance for Features:

	Internal Feature	External Feature
MMC	Actual produced size - Maximum Material Condition = Bonus Tolerance.	Maximum Material Condition – Actual Produced Size = Bonus Tolerance.

Total Tolerance = Tolerance from feature Control Frame + Bonus

Tolerance at MMC:

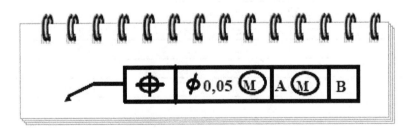

An example of a feature allowing bonus tolerance is shown in the following figure. The through hole's size is given as limits, with 12-12.5 allowable. The perpendicularity requirement of the hole relative to datum A is given by the feature control frame. The

tolerance of 0.2 applies only in the case when the hole is produced at its maximum material condition of 12 millimeters. For a hole of any other size, bonus tolerance is available. The amount of bonus tolerance, and therefore the total tolerance on perpendicularity, depends on the actual produced size of the hole.

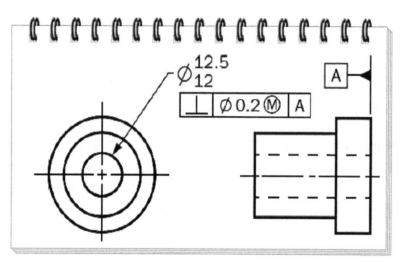

The next table illustrates the bonus tolerance calculation and the total tolerance on the perpendicularity of the feature for various possible hole sizes that might be produced. The table does not show all possible hole sizes, only sizes in 0.1-millimeter increments. For any other size, the calculation is used.

Bonus Tolerance Calculation for Internal Feature, MMC Basis:

	Hole, Actual Produced Size		Maximum Material Condition		Bonus Tolerance		Tolerance from Feature Control Frame		Total Tolerance
	12.5	-	12	=	0.5	+	0.2	=	0.7
LMC:	12.4	-	12	=	0.4	+	0.2	=	0.6
	12.3	-	12	=	0.3	+	0.2	=	0.5
	12.2	-	12	=	0.2	+	0.2	=	0.4
	12.1	-	12	=	0.1	+	0.2	=	0.3
MMC:	12	-	12	=	0	+	0.2	=	0.2

Tolerance at LMC:

With the LMC modifier, the geometric tolerance given on the drawing applies when the size of the feature actually produced is its least material condition; for a feature produced at any other size, there is bonus tolerance. An example of a feature allowing bonus tolerance is shown in the following image.

The through hole's size is given by the size tolerance, and a feature control frame indicates the position requirement of the hole relative to datum's A and B; see the next image. The position tolerance of .006 applies only in the case when the hole is produced at its least material condition of 1.502 inches.

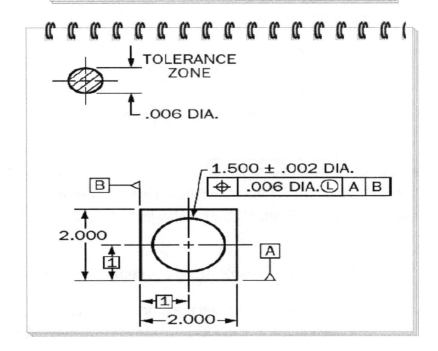

CHAPTER 8

Features, Feature Control Frames and Datum's:

Remember that a feature is a general term applied to a physical portion of a part such as a surface, hole or slot. A feature may be considered a part surface. Parts are made up of features, which can be divided into two classes: features that have size and surfaces. Successfully interpreting prints with GD&T requires you to think about your parts in terms of their features, not only in terms of the dimensions appearing on the print. Features that have size have a directly dimensioned size tolerance, which may be placed explicitly next to the dimension or implicitly on a title block or symbolic key. The maximum and minimum limits of size are given special names in the geometric system, which are helpful for understanding bonus tolerance.

Parts are made up of features, which can be divided into two classes: features that have size and features that are surfaces. Successfully interpreting prints with GD&T requires you to think about your parts in terms of their features, not only in terms of the dimensions appearing on the print.

Features that have size have a directly dimensioned size tolerance, which may be placed explicitly next to the dimension or implicitly on a title block or symbolic key. The maximum and minimum limits of size are given special names in the geometric system, which are helpful for understanding bonus tolerance:

Maximum material condition (MMC):

The maximum size limit for an external feature or the minimum size limit for an internal future. In either case, whatever limit, if produced, results in the part having the maximum amount of material allowed.

On the other hand, the Least Material Condition (LMC), is the minimum size limit for an external feature or the maximum size limit for an internal future. In either case, whatever the limit, if the part is produced, the end result is the part having the least amount of material allowed.

The terms "MMC" and "LMC" are useful for calculating clearances between mating parts.

A fundamental rule of geometric dimensioning and tolerancing is that a size dimension controls not only the size of the feature, but also its form. Any feature with size that is produced at its MMC is only acceptable if its form is perfect.

The size of the pin in the figure below is allowed to vary from 15.98 to 16. Variation from perfect form is only allowed to the extent that the size varies from MMC.

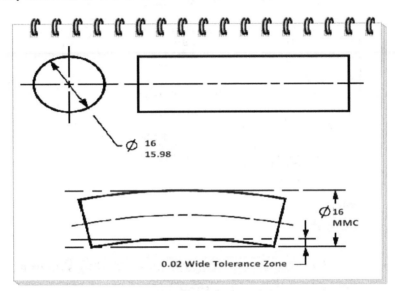

Some form variation is allowed (bow, necking, barreling, out-of-round, etc.), but only variation that does not push the material of the pin beyond its 16-millimeter cylindrical boundary.

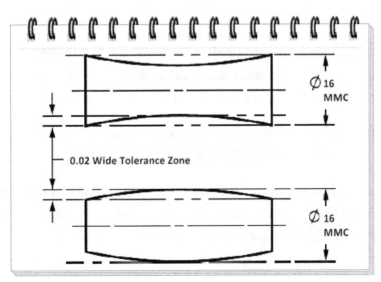

- Are comprised of part surfaces, or elements, that are internal part surfaces such as a hole diameter or the width of a slot.

- A feature of size dimension is a dimension that is associated with a feature of size. A non-feature of size dimension is a dimension not associated with a feature of size.

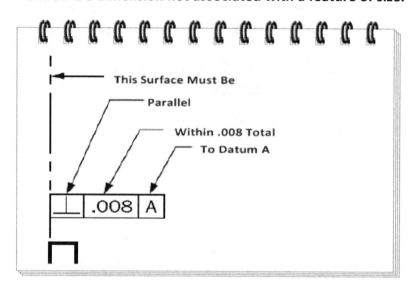

The rectangular box is called a feature control frame and indicates a geometric tolerance. The proper way to read a feature control frame is shown above.

A feature control frame consists of a series of symbols and numbers. From left to right:

1. The geometric characteristic symbol: This symbol indicates what type of geometric tolerance applies.

2. The shape of the tolerance zone (optional): The shape will be identified by one of the following: square symbol, diameter symbol, spherical diameter (rarely), or blank. When no symbol is given, the shape of the tolerance zone is inferred from the shape of the feature.

3. The total size of the tolerance zone: This is always given in the same linear dimension units used on the rest of the print—inches or millimeters.

4. The material modifier symbol (optional): The symbol will be one of these three: the MMC symbol, the LMC symbol, or the RFS symbol (prior practice). For MMC, bonus tolerance is available on an MMC basis; and for LMC, bonus tolerance is available on an LMC basis. If no symbol is given or if the RFS symbol is used, there is no bonus tolerance allowed.

5. The datum reference frame (mandatory for some geometric characteristics, excluded for some, optional for some): This will be one or more datum references listed in order of precedence (primary, secondary, etc.).

The Apprentices Guide to Blueprint Reading

An Example of a Feature Control Frame:

If a hole, for instance, has the following size and geometric control, and the hole measures .502. It would be incorrect to use a bonus tolerance of .003 (.502 - .499(MMC)) if the hole is not perfectly oriented to the Datum's. If the hole is out of perpendicular to datum A by .002, for instance, the bonus that may be used is reduced by that amount. The bonus would be merely .001 and the allowable position tolerance = .016.

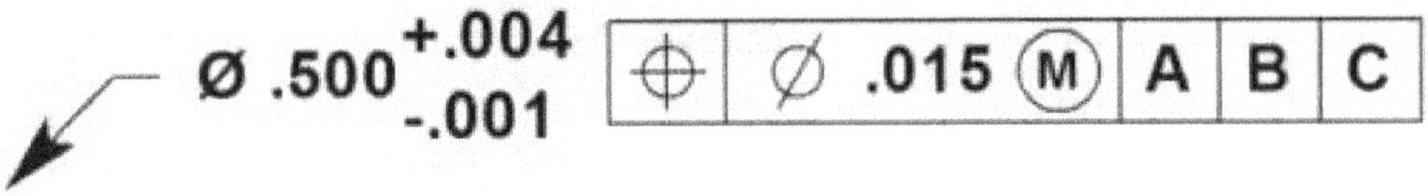

$\varnothing .500^{+.004}_{-.001}$ | ⌖ | $\varnothing$ .015 Ⓜ | A | B | C |

Table A-2

Symbol	Meaning
Ⓜ	Maximum Material Condition (MMC)
Ⓛ	Least Material Condition (LMC)
NONE	Regardless of Feature Size (RFS) Current Practice.
Ⓢ	Regardless of Feature Size (RFS) Old Style
▷	Translation
Ⓟ	Projected Tolerance Zone
Ⓕ	Free State

Symbol	Meaning
Ⓤ	Unequally Disposed Profile
Ⓘ	Interdependency
Ⓣ	Tangent Plane
⟨ST⟩	Statistical Tolerance
⟨CF⟩	Continuous Feature
CR	Controlled Radius
□	Square
[A]◄	Datum Feature Identifier

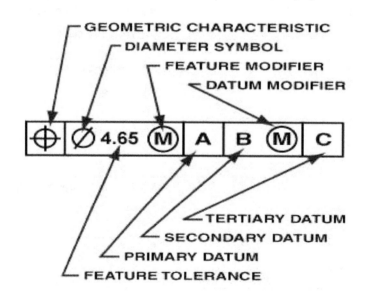

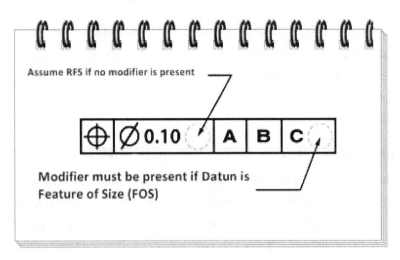

DATUMS:

A datum very important concept in the geometric system (GD&T). A datum is a theoretical point, axis or plane derived from the true geometric counterpart of a specified datum feature. A datum is a perfectly flat surface or perfectly straight axis represented by the physical inspection equipment.

A datum serves as the origin of all measurements in the geometric system and provides the way to locate and orient the part in its inspection setup.

The datum feature symbol shows how to indicate a feature that is to be used as a datum for controlling other features.

Datum A is the primary, B is the secondary, and C is the tertiary datum. In this picture it appears that each datum is a planed surface, but this is not always the case. Technically, it is never the case. This is because a perfectly planed surface is theoretical and can never be achieved in practice. Once this premise is accepted, a deeper understanding of locating principles can be achieved.

Remember that three points define a plane and two points define a line. After the primary datum is located using three points, the secondary datum will typically use two points to define a line. The only degree of freedom remaining can be constrained by a single point on the tertiary (third) datum. Therefore in a standard datuming scheme, 6 points fully locate a part or assembly; 3 for the primary, 2 for the secondary, and 1 for the tertiary datums.

DATUM A:

Take a look at the primary locating datum which is shown as a plane. Let's imagine this plane as a nice flat granite surface. Once the part is placed on the granite surface, only three points of it will be actually contacting the granite. Remembering that no surface can be perfectly flat (the granite nor the part), means only the lowest three points of the part will be touching the highest three points of the granite where these intersect. Since no two parts are identical these three exact points that touch the granite, and therefore locate the part, will not be the same from part to part.

If necessary, there are ways to control what locations on the parts surface are used to create the reference coordinate system. Instead of using a planer surface as the locating feature, three distinct locators can be used. By using three 'point' locators, the same three areas of every part are used to locate it. This can reduce the amount of locating tolerance, and improve down-stream fit-up or function depending on where the point locators are placed. Three 'point' locating targets are used to create the primary datum for a component.

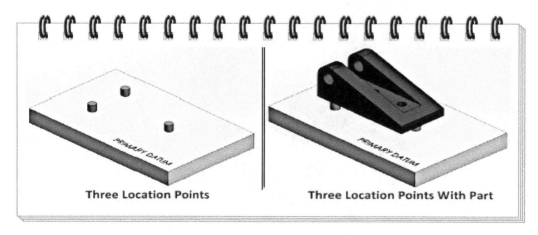

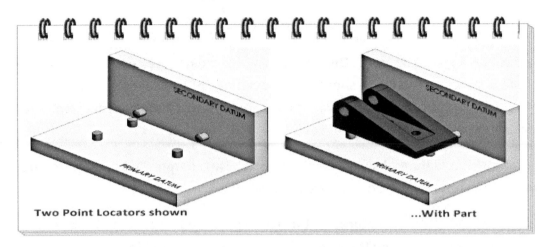

The locators are considered points because the top of the locators are spherically shaped, so the center is the highest point. Once the B and C datums are added, the primary datum plane formed by this fixture will repeatedly locate off the same three points on the parts. Instead of relying on the lowest three points of a part, wherever they may be, this configuration dictates the location of these three points. Notice how far apart the locators are. If the three points are moved to be very close to each other then you can imagine the part would become unstable, therefore not as repeatable. This principle applies to all part locators; farther apart is better.

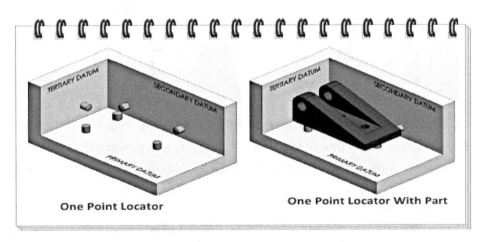

DATUM B:

Similar principles apply to datum B. A planer surface can be used to locate datum B, which means two *undefined* points will be touching the planer surface. In some cases this may work fine. When more control is needed, two distinct point locators may be used.

DATUM C:

Similar principles apply to datum C. Once a reference coordinate system is created with Datums, the dimensioning scheme should take advantage of this new coordinate system. Without extenuating circumstances, dimensions should go to a Datum's edge, which produce more consistent parts for no additional cost.

A datum feature is a feature of a part, is used to establish a datum. Datum features are indicated on the print using the datum feature identifier.

Datum's are derived from the features on the actual produced part, called "datum features." Datum features are indicated on the print using the datum feature identifier.

Datum Example:

The symbol or Datum feature identifier is shown in the image above, the Datum symbol is attached to the datum feature, or a leader line from the feature. Note that datum symbols are identified by letter codes starting with the letter A, following the order of the alphabet, A,B,C, etc.

It is important to know that more than one datum can be used to control a feature, as can be seen in the following image. Datum's are needed any time there is a need to specify the relationship of one feature to another.

Datum's are needed any time you have to specify the relationship of one feature, to another.

Datum's are not referenced for size tolerances or form controls, but they are needed for most other geometric controls.

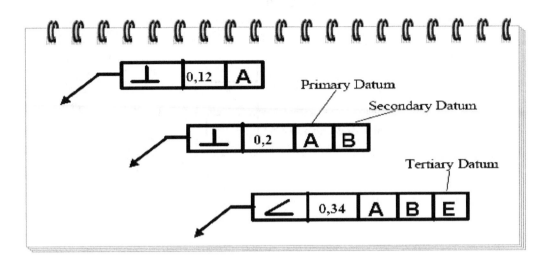

Example terms and symbols:

Datums are the geometric counterpart to their datum features. A feature intended to be a flat plane means the corresponding datum is a plane. Several examples of flat features used for datum planes can be found in the next image.

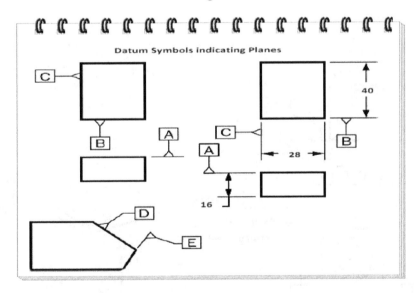

A round feature such as a shaft used as a datum feature results in a datum axis. The figure below shows this.

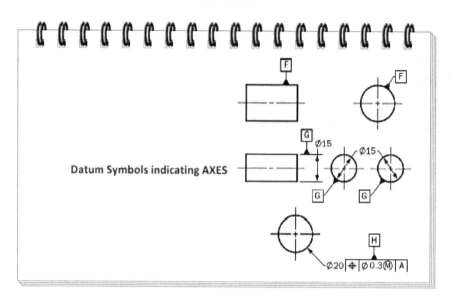

Usually several datums are indicated on a print. The design intent of the part is that datum features D, E, and F are all flat surfaces perpendicular to each other, so datums D, E, and F are flat surfaces perpendicular to each other. Datums are needed any time there is a need to specify the relationship of one feature with respect to another.

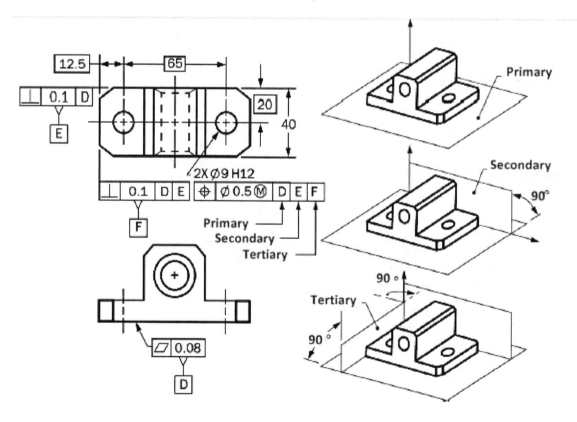

In the above drawing, it is important to locate the two holes relative to the datum features D, E, and F. Remember that Datums are not used for size tolerances or form controls, but they are needed for most other geometric controls.

The 3-2-1 Datum System is shown in the image below. Note that the datum planes are perpendicular to each other.

The primary datum A provides 3 points of contact to locate a part, it restrains three degrees of freedom, 2 rotation and 1 translation. The primary datum locates a plane.

• The secondary datum B provides 2 points of contact, it restricts 2 degrees of freedom, 1 rotation and 1 translation, it locates a line.

• The tertiary datum C provides 1 point of contact, it restrains 1 degree of freedom (Translation). It locates a point.

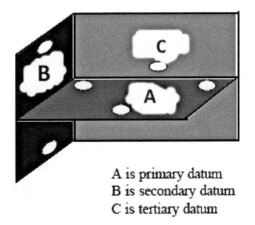

A is primary datum
B is secondary datum
C is tertiary datum

3-2-1 Datum System

So in total the three datums restrain six degrees of freedom, which is sufficient to locate a part.

CHAPTER 9

Form, Form Controls and Geometric Tolerances:

A description of a Form control hole at LMC and its location

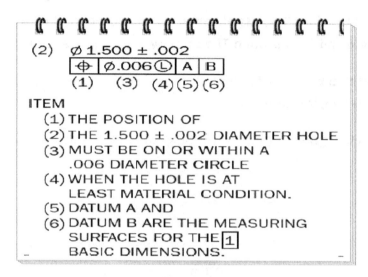

Form controls limit the variation a feature is allowed to have from its perfect shape, or form. Form controls do not specify the size of features or any other type of variation. Because the size tolerance controls the form of a size feature, a geometric form control applied to a size feature tightens the form control.

• A form tolerance states how far an actual surface or feature is permitted to deviate from the desired form as specified in a drawing. This includes flatness, straightness, circularity, and cylindricity, the profile of a surface and the profile of a line.

In GD&T, there are four tools for controlling form: straightness, flatness, circularity, and cylindricity and their symbols are shown below.

• **An example for straightness:**

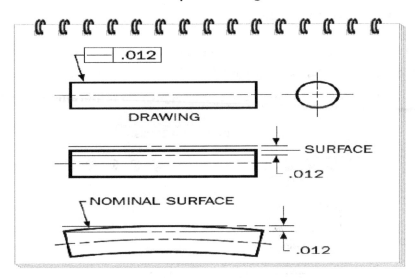

NOTE: All points on the surface must be between two parallel lines that are .012 apart and share a common plane with the nominal surface.

Straightness and flatness are the two most important types of geometric controls. This is true because before a feature can be used as a datum it must be straight or flat. Straightness controls the lines drawn on a surface, the axes of shafts and holes and the edges of parts. The image below shows a straightness control for lines on a surface. The tolerance zone is specified by two parallel straight lines separated by a tolerance value. Notice the lines are drawn in a plane parallel to the plane of projection. Each line drawn on the surface must lie within the tolerance zone. The lines are drawn in the direction as shown and the call out applies to the entire surface.

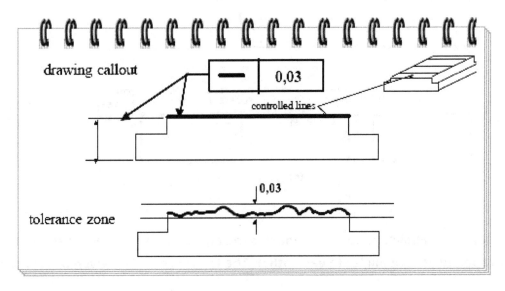

It would not be cost effective to inspect the entire surface, instead 3 or 4 spaced lines are checked to insure the parts compliance.

It is important to note that the tolerance zone applies to each line independently. The tolerance zones for each line may vary (up or down) relative to each other. The surface could be wavy, convex, and concave and still meet the straightness requirement.

• An example of Flatness:

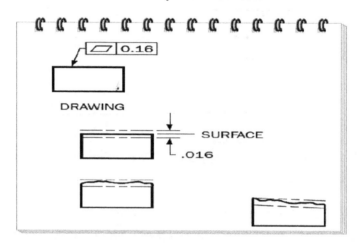

NOTE: All of the points of the surface must lie between two parallel lines .016 apart.

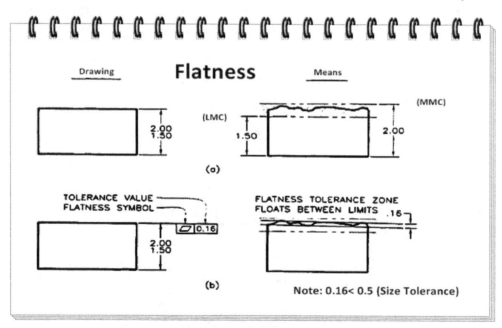

A flatness tolerance applies to surfaces, and it is the equivalent to two equal straightness tolerances applied at right angles to each other. The tolerance zone is given by two parallel planes whose distance apart is equal to the tolerance zone.

Note: All points on the surface must lie within the tolerance zone. Flatness Control, controls deviations such as waviness, concavity and convexity.

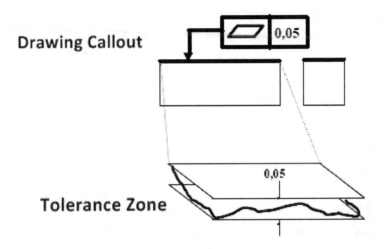

Drawing Callout

Tolerance Zone

The measuring principal is the same as it is for straight lines on planes, except that it is now necessary to correct for inclination in two directions.

Straightness:

Straightness controls lines drawn on a surface, the axes of shafts and holes and the edges of parts. The drawing callout is applied to the view, which indicates the profile of the feature to be controlled. The image below shows a straightness control for lines on a surface. The tolerance zone is defined by two parallel straight lines whose distance apart is given by the tolerance value 0,02. The lines are drawn in a plane parallel to the plane of projection in which the feature is indicated. The actual position of the tolerance zone relative to other features, such as surface A, is not controlled.

Each line drawn on the surface must lie within the 0,02 tolerance zone. The lines are drawn in the direction as shown. The callout applies to the entire surface. It is not cost effective to inspect the entire surface, so instead 3 or 4 lines spaced apart would be checked to make sure that the part complies with the print.

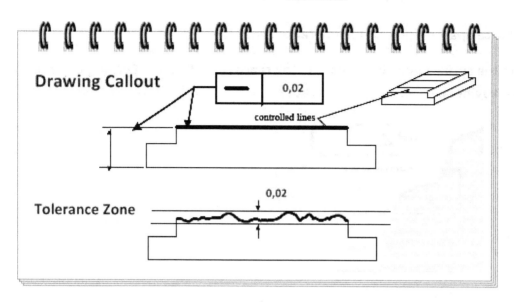

Note that the tolerance zone applies to each line on the surface independently. The tolerance zones for different lines may vary in location, up or down the surface relative to each other. Meaning that a surface could be wavy, convex or concave and still meet the straightness requirement.

Measuring the Straightness of a Plane Surface:

The image below illustrates the measuring technique. Note that the surface being measured may be inclined relative to the indicator. If no correction is made for the slope, then the measured tolerance will be larger than the actual value. The image shows how to correct for slope by plotting the results on graph paper.

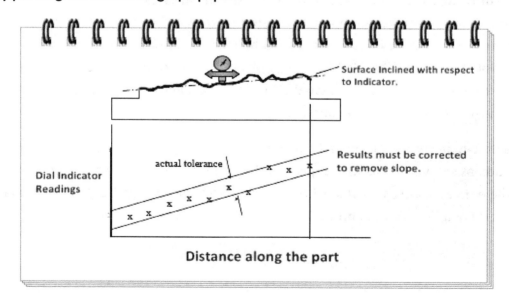

Another technique is to zero the indicator at three well spaced points along the surface being measured. In order to use this technique, the bottom surface of the part has to be supported by adjustable supports that can be adjusted to give zero readings. Once the adjustments are made, the indicator readings can be used to determine the tolerance. The tolerance is the distance between the highest and lowest readings.

CIRCULARITY:

A circularity tolerance is used to control the roundness of a feature, e.g, the circumference of a shaft or a hole. The tolerance zone applies to a line drawn around the circumference of the part at a cross section that is at right angles to the plane of projection. Each line must lie within the tolerance zone, Note that the tolerance zones are independent of each other. The tolerance zone is defined by two concentric circles set apart by a radial distance equal to the tolerance value. (In the Image below 0,02) Note that the position of the tolerance zone relative to the axis of the part is not controlled.

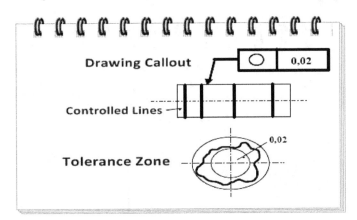

- An example of Circularity:

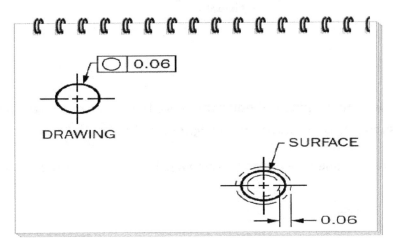

NOTE: According to the above image, all points on the surface must lie between two concentric circles that are 0.06 apart on the radius.

The image below shows an approximate method for checking circularity, note that the part is rotated while the dial indicator remains stationary. The total indicator movement (TIM) is noted (The difference between the highest and lowest indicator readings). Half of this value is the deviation.

- TIM=max reading-min. reading
- Deviation=TIM divided by 2

•Remember use two points of measurement a 180 degrees apart.

This method of measurement will not detect lobing. Lobbing occurs from machining operations such as centerless grinding and reaming. If lobing is suspected other methods of, inspection should be used.

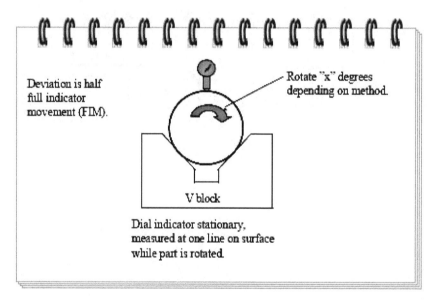

For cylindrical shapes, shafts, pins and holes there are two types of straightness controls. The following image shows how to control the straightness of the outside surface.

Note that the tolerance zone is depicted by two lines in a plane thru the center of the part.

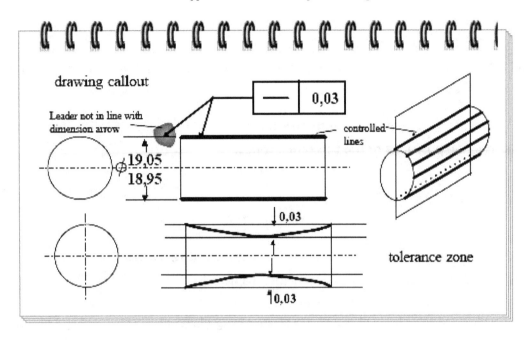

Looking at the image above, all of the actual circular elements of the surface must be within the specified tolerance 18, 95-19, 05. As well as the boundary of perfect form at MMC 19, 05. It is important to note that each longtitudinal line of the surface must lie in the tolerance zone as specified by the two parallel lines with a distance equal to the straightness tolerance 0, 03.

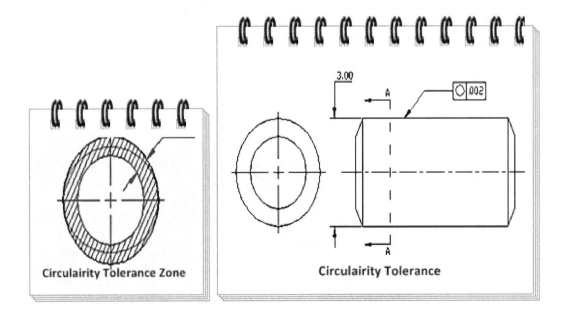

- The callout applies to the entire surface, but in practice several lines would be checked in order to verify that the part is in compliance.

- The part (shaft) may be bowed or bent and still meet its requirement, because it does not control the relative positions of the tolerance zones.

CYLINDRICITY:

Cylindricity controls the roundness of a feature over its entire surface. The tolerance zone is a tube whose thickness is given by the tolerance value. The zone is formed by the 360° rotation of a cylinder whose diameter is equal to the tolerance value and whose length is equal to the length of the feature being controlled. The diameter symbol is used for this tolerance. Note that Cylindricity is checked like Circularity.

All points on the surface must lie within the tolerance zone, so cylindricity controls deviations such as concavity and convexity.

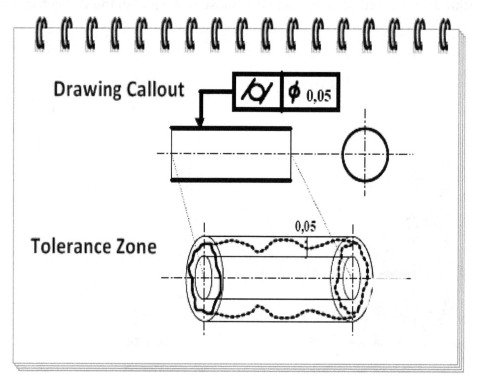

• **An example of Cylindricity:**

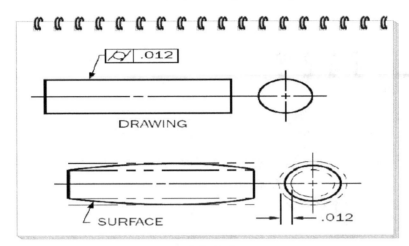

NOTE: According to this drawing, all points on the surface must lie between two concentric cylinders that are .012 apart on the radius.

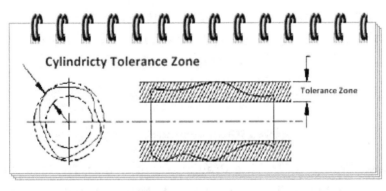

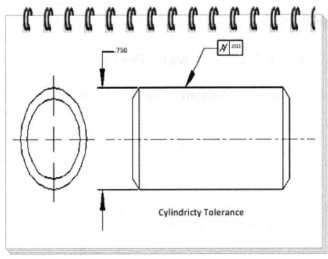

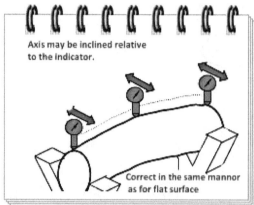

Measuring the straightness of a cylindrical surface is a bit more complicated than for a flat surface. V-blocks can be used to support the part; however, V-blocks can add additional error to the part. Notice the dial indicator is moved across the part at each measuring point. The technique used is the same as for a flat surface. Common practice is to measure four times around the part with a spacing of 90 deg. All measured sections must reside within the straightness tolerance.

FORM TOLERANCE FOR LINES

- The Straightness of an axis:

The next image shows how to control the straightness of an axis. The leader for the Feature Control Frame is attached to the dimension line. Note that a Diametral symbol is added to the feature control frame to show that the tolerance zone is a cylinder.

- Straightness Tolerance of an axis can fall under RFS or MMC conditions.

RFS: Each actual local size or circular element must be within the specified size tolerance (18,95-19,05).

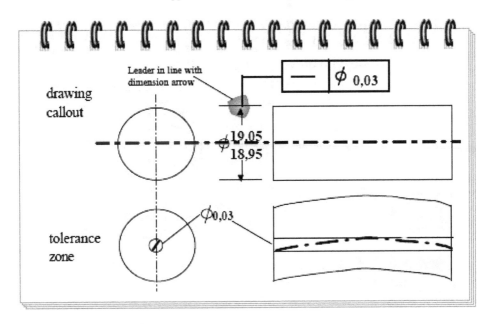

RFS: The derived median line (estimated actual axis) must lie within the tolerance zone of 0,03 regardless of the size of feature (regardless of the diameter of the part). The same tolerance applies to parts at MMC (19,05) and LMC (18,95) and for all sizes in between.

MMC: Each actual local size (Circular element) must be within the specified size tolerance (18,95-19,05). The derived median line (estimated actual axis) must lie within a tolerance zone of 0,03 at MMC. If the part is smaller than MMC then the difference in size from MMC can be used to increase the straightness tolerance. So the straightness tolerance is different for different size parts. At LMC the straightness tolerance is 0,03 + 0,1= 0,13.

MMC is used for the functional design of assemblies (Mating Parts) and allows for functional gauging, reducing manufacturing and inspection costs.

- Measuring straightness of an axis – RFS:

To measure the straightness of an axis at RFS two dial indicators are needed. The part is held between centers. In each longitudinal section the value R=(Au-A1)/2 are determined. Where Au is the reading of the upper indicator and A1 is the reading of the lower indicator. The difference between R max and R min. within one section represents the straightness deviation of the axis for this section. The straightness deviation of the axis is the maximum of the section deviations, Note at least 4 sections must be measured. The method shown in the image below is an approximate method for measuring straightness.

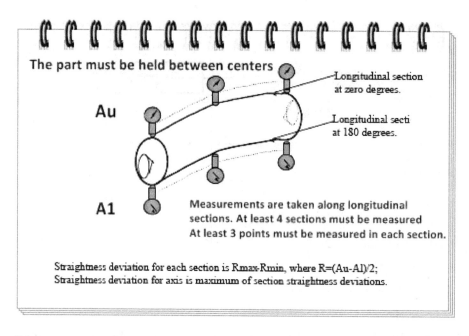

It is not possible to measure straightness of an axis directly. The only measurements that can be taken are those on the outside surface of the part. This is true for all measuring methods.

In addition to meeting the straightness requirements, circular elements along the part must meet local size feature requirements. (According to the first image they must lie between 18,95 and 19,05)

The next image shows some examples of measurements for different shaped parts.

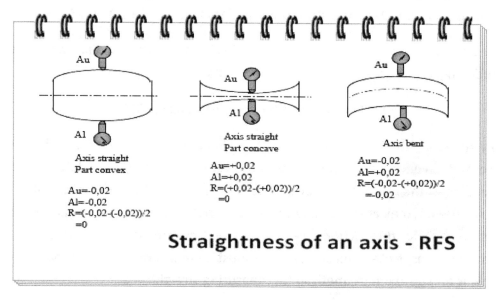

The above image shows how straightness is calculated. If the part has a straight axis and is symmetrical about the centerline, then the upper and lower indicator readings cancel each other out.

If the axis is bent, not straight, then one indicator will read lower than the other. Each indicator is measuring the error so the sum of the two readings is twice the amount of actual error. The actual error is calculated by dividing the sum by two. This gives you an average value estimate for the position of the axis.

• PROFILE CONTROLS •

A profile is the outline of an object in a given plane (two-dimensional figure). Profiles are formed by projecting a three-dimensional figure onto a plane or by taking cross sections through the figure. The elements of a profile are straight lines, arcs, and other curved lines. If the drawing specifies individual tolerances for the elements or points of a profile, these elements or points must be individually verified.

With profile tolerancing, the true profile may be defined by basic radii, basic angular dimensions, basic coordinate dimensions, basic size dimensions, un-dimensioned drawings, or formulas.

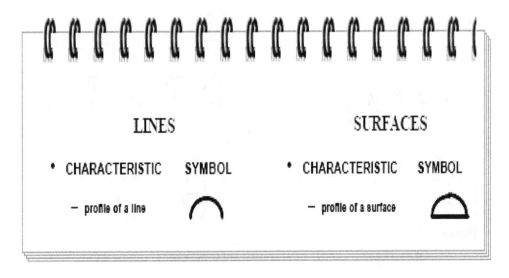

Profile controls are applied only to surfaces and limit all types of variation: location, orientation, and form. They can be used for simple flat, inclined, and curved surfaces as well as more complex 3D surfaces. The shape of the tolerance zone follows the basic shape of the surface(s) controlled and is equally disposed bilaterally by default. In most cases profile feature control frames reference at least one datum, usually more.

In the geometric system, there are two profile controls: profile of a surface and profile of a line. Profile of a surface is a full 3D control, and profile of a line is a 2D control applied only at individual cross sections. Examples are shown below.

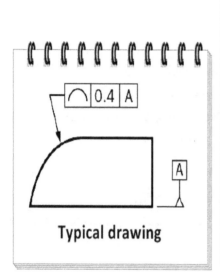

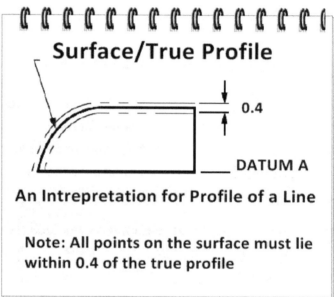

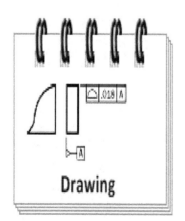

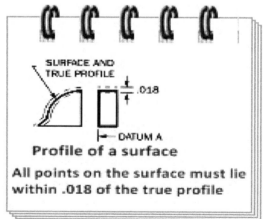

(An example of the profile of a line control)

•ORIENTATION CONTROLS•

An orientation tolerance specifies a zone within which the considered feature, its line elements, its axis, or its center plane must be contained.

Orientation Tolerances are Angularity, parallelism, perpendicularity, and in some instances, profile can be an orientation tolerance as well. They are applicable to related features. These tolerances control the orientation of features to one another.

An orientation deviation, is a deviation from nominal form and orientation. Orientation deviations are produced by the same machining conditions that produce form deviations. They can also be caused by setup errors when the machining datum's are changed.

Orientation Controls	//	Parallelism
	⊥	Perpendicularity
	∠	Angularity

Orientation controls limit the tilt of features and are always associated with basic angle dimensions. For parallel and perpendicular angles, these angular dimensions often do not actually appear on the print since they are usually implied. Because they control the relationship between features, orientation feature control frames always reference at least one datum.

Orientation controls are used for features of size and surfaces. When applied to surfaces, they also control the form. They do not control the location of features. When applied to features of size, the center plane or axis of the feature is controlled, and bonus tolerance is available.

Orientation controls provide full 3D control, but they can be made 2D by the note "LINE ELEMENTS" appearing beneath the feature control frame. In the geometric system, there are three tools for controlling the orientation of features: parallelism, perpendicularity, and angularity. Examples can be seen below.

ORIENTATION CONTROL FOR PARALLELISM:

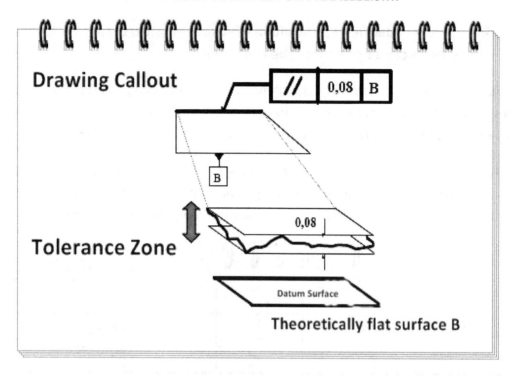

(Example of Parallelism)

Looking at the above image we see that an orientation tolerance is specified relative to one or more datum's. In this image the tolerance zone is defined relative to datum B, which is the bottom of the part.

The tolerance zone in the above image is defined as two plane surfaces parallel to datum B. All of the points on the surface that are being controlled, must lie within the tolerance zone. Note: the tolerance zone controls the orientation of the surface, relative to datum B, but not its vertical position.

A few more examples of orientation Controls at RFS and MMC follow.

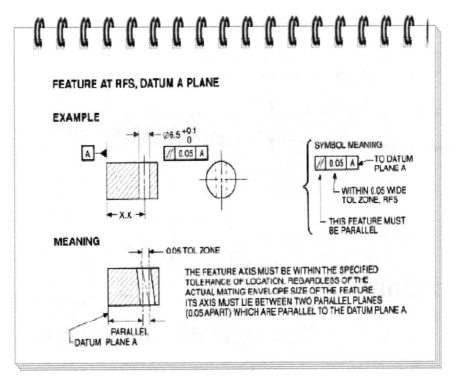

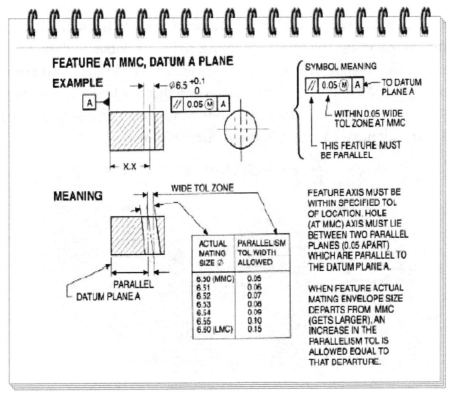

ORIENTATION CONTROL FOR PERPENDICULARITY:

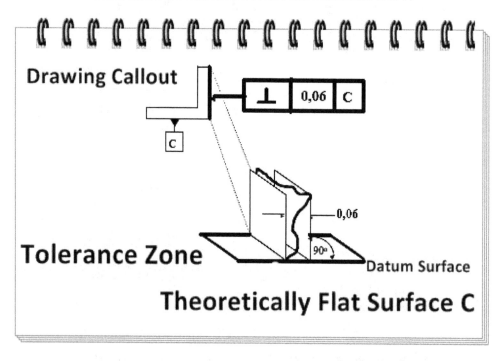

Looking at the above image we see that an orientation tolerance is specified relative to one or more datum's. In this image the tolerance zone is defined relative to datum C, which is the bottom of the part. The tolerance zone in the above image is defined as two plane surfaces at right angles to datum C. All of the points on the surface that are being controlled, must lie within the tolerance zone. Note: the tolerance zone controls the orientation of the surface, relative to datum C, but not its position in the horizontal direction.

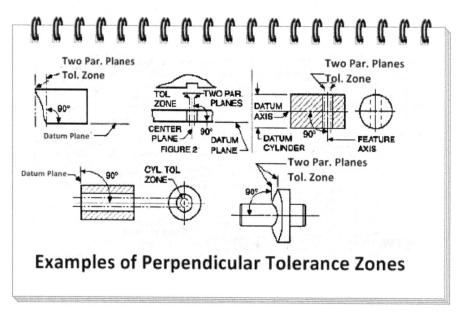

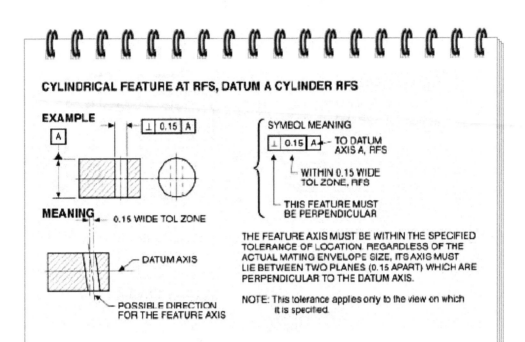

Example:

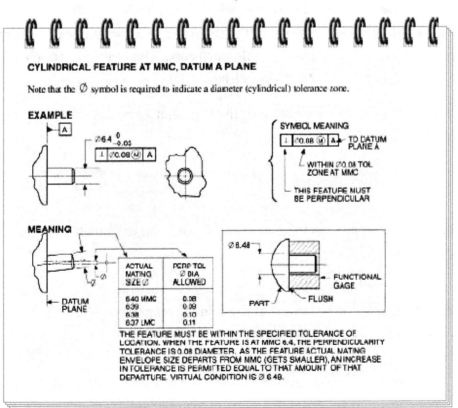

ORIENTATION CONTROL FOR ANGULARITY:

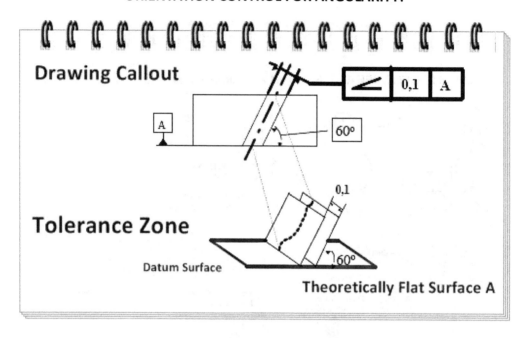

Note: An orientation tolerance is always specified relative to one or more datum's. The angle of an orientation tolerance can be any angle except 0° or 90°. Looking at the image above the tolerance zone is defined relative to datum A, which is the bottom of the part. The tolerance zone is defined by two plane surfaces set at the theoretically exact angle 60° to datum A. All points on the hole axis must lie within the tolerance zone. The tolerance zone could also apply to an axis, in which case it may be cylindrical.

The tolerance zone defines the orientation of the axis relative to datum A, but not its position in the horizontal direction.

Question: How do we know if a tolerance applies to an axis or a surface for cylindrical shapes?

Answer: If the tolerance applies to the axis then a leader line will be attached to the dimensioning line.

Note that tolerances may apply to features that vary in size. When this occurs the designer needs to specify the size of the feature that the tolerance applies. (See MMC) In the above image the hole diameter can vary. The variation will affect how the angularity is measured.

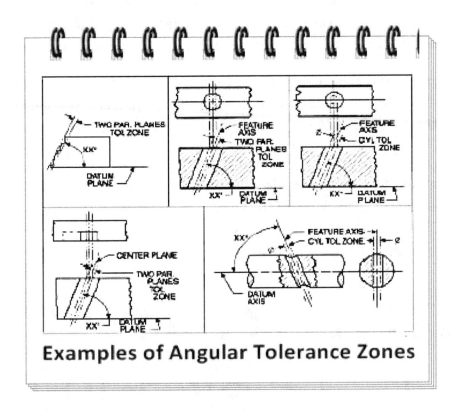

Examples of Angular Tolerance Zones

• LOCATION CONTROLS •

A location tolerance controls position as well as orientation and form. So for a hole, a location tolerance will control the position of the hole axis relative to the specified datum's, the tilt of the axis of the hole, and any form deviations, like convexity or concavity of the axis.

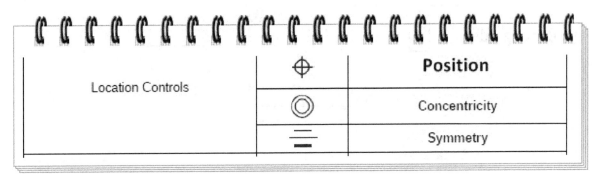

Location tolerances control position as well as orientation and form. As an example a location tolerance will control the position of a hole axis, relative to the specified datum's, the tilt of the axis of the hole, as well as convexity and concavity of the axis.

A position tolerance controls the position of a feature relative to one or more datum's. The image below illustrates how the axis of a hole is controlled. The 3-2-1 datum system is used with A as the primary datum, B as the secondary datum and C as the tertiary datum.

Note the tolerance zone is a cylinder set at right angles to the primary datum with a diameter equal to the tolerance value. The location of the axis of this cylinder is specified by theoretically exact distances of 65mm. and 30mm. from the secondary and tertiary datum surfaces.

All points on the axis of the hole must lie within the tolerance zone. Note that the tolerance zone applies over the complete depth of the hole.

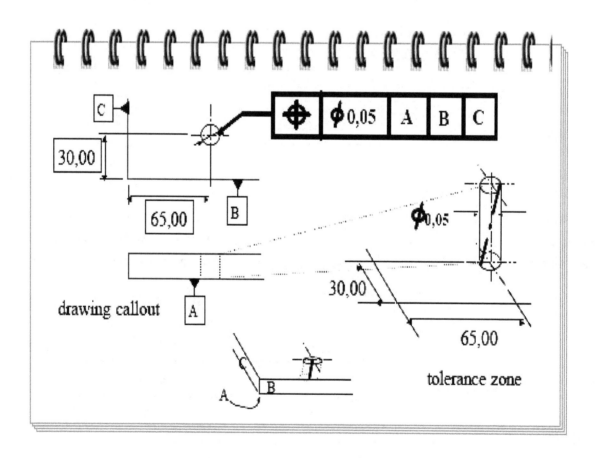

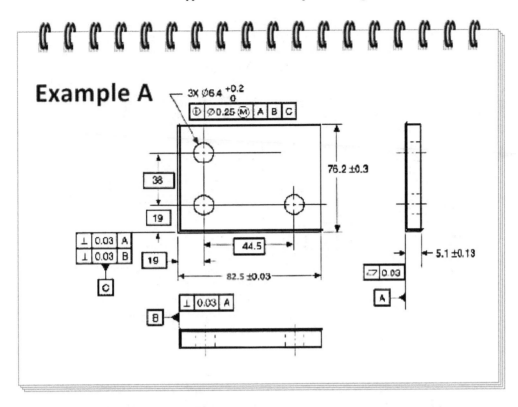

The above image and the one below are examples of hole positioning. They depict how the datum's are controlled relative to each other. A is the primary datum, B is the secondary datum and C is the tertiary datum.

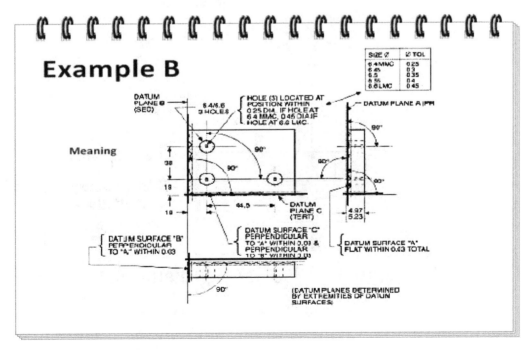

• POSITION CONTROLS •

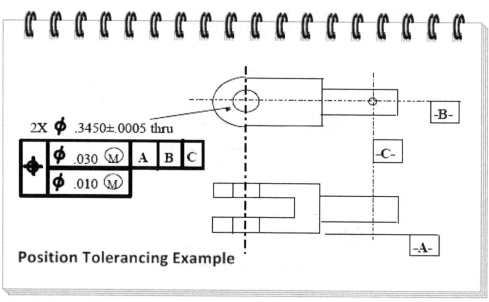

Position Tolerancing Example

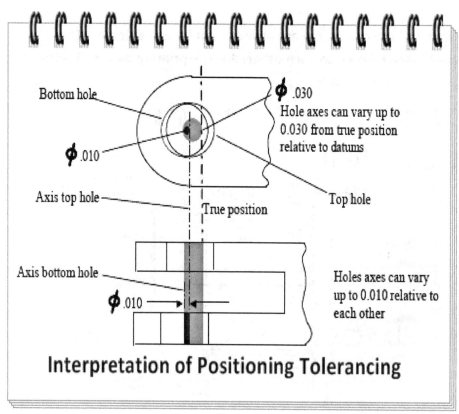

Interpretation of Positioning Tolerancing

• CONCENRICTY CONTROLS •

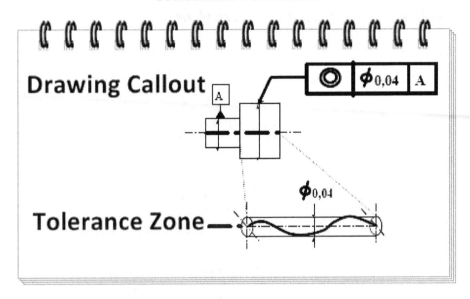

A concentricity or coaxiality tolerance, controls the axis of one feature relative to the axis of another feature. Using the image above as an example, the axis of the larger cylinder must lie within a cylindrical tolerance of 0,04 mm. of the axis of the smaller cylinder, datum A. Note the tolerance zone applies over the full length of the feature being controlled.

• Symmetry Control •

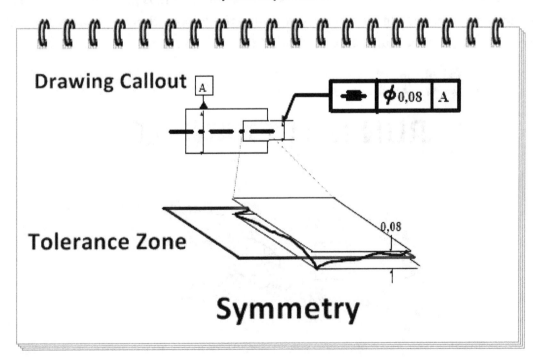

The median face should lie between the two parallel planes set 0,08 mm. apart. They are symmetrically positioned about the datum median plane.

Note: Datum A is at RFS.

• RUNOUT •

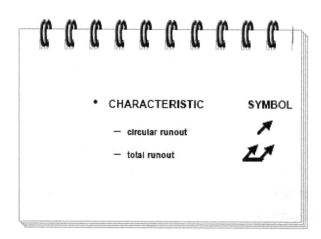

A run-out tolerance specifies how far an actual surface or feature is permitted to deviate from the desired form depicted in a blueprint during the full rotation of the part on a datum axis. There are two types of run-out, Circular and Total run-out.

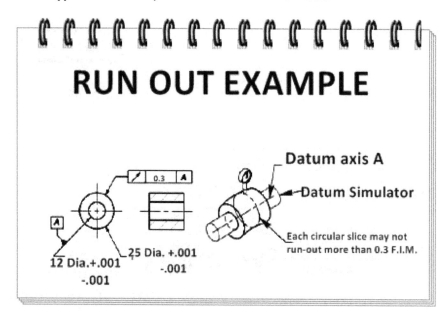

Circular Run-Out

- A dial indicator is often used to verify a runout control
- First, the part is located in a chuck or collet to establish datum axis A.
- A dial indicator is placed on the surface being checked.
- As the part is rotated 360 degree, the dial indicator movement is the run out value of the circular element.
- Several independent dial indicator readings are made at different places along the diameter.

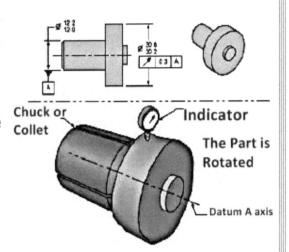

Total Run-out

- Total run out is used to control the combined variations of circularity, straightness, coaxiality, angularity, taper and profile when applied to surfaces around and at right angles to a datum axis.

- Note that total runout cannot be applied to conical or curved surfaces as can circular runout.

LIMIT TOLERANCE:

When a dimension has a high and low limit stated. 1.250/1.150 is a limit tolerance

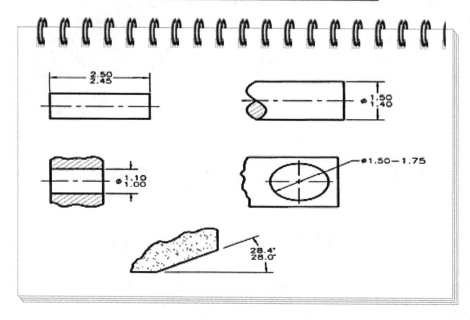

PLUS or MINUS TOLERANCE:

The nominal or target value of the dimension is given first, followed by a plus-minus expression of tolerance. .32 +/- .004 is a plus-minus tolerance.

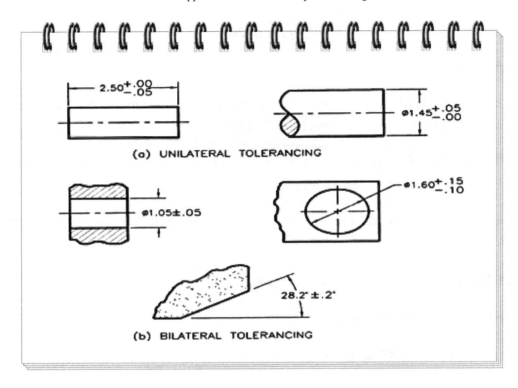

ALLOWANCES:

Allowance is an intentional difference in dimensions of mating parts to provide the desired fit. A clearance allowance permits movement between mating parts when assembled.

ALLOWENCE = MMC Hole – MMC Shaft

For example, when a hole with a 0.250-inch diameter is fitted with a shaft that has a 0.245-inch diameter, the clearance allowance is 0.005 inch. An interference allowance is the opposite of a clearance allowance. The difference in dimensions in this case provides a tight fit. Force is required when assembling parts, which have an interference allowance. If a shaft with a 0.251- inch diameter is fitted in the hole identified in the preceding example, the difference between the dimensions will give an interference allowance of 0.001 inch. As the shaft is larger than the hole, force is necessary to assemble the parts.

CLEARANCE:

Clearance is defined as the loosest fit or maximum intended difference between mating parts.

CLEARANCE = LMC Hole- LMC Shaft.

Types of Fit:

Standard ANSI Fits:

Running and Sliding fits (RC) are intended to provide a running a running performance with a suitable lubrication allowance. It ranges from RC1 to RC9.

Force fits (FN) or shrink fits are special types of interference fits. The fit requires constant pressure. Force fits range from FN1 to FN5.

Force fits are also known as Interference, or shrink fits. The calculation for the smallest amount of interference is:

Min. Interference = LMC shaft – LMC hole

The largest amount of interference is:

Max Interference = MMC shaft – MMC hole

Locational fits determine only the location of the mating parts.

Clearance fit:
- The parts are toleranced so that the largest shaft is smaller than the smallest hole.
- The allowance is positive and greater than zero.

Interference fit:
- The Max. clearance is always negative.
- The parts must always be forced together.

Transition Fit:
- The parts are toleranced so that the allowance is negative and the max. clearance is positive.
- The parts may be loose or forced together.

One cylindrical or spherical surface or a set of two opposed elements or opposed parallel surfaces associated with a size dimension. Geometric tolerancing is a three-dimensional (3D) language. The geometric tolerance creates a zone of tolerance within which the feature must lie. Tolerance zones come in a variety of shapes depending on the control applied and the feature.

Tolerance zones can be three-dimensional or two-dimensional. Two dimensional (2D) tolerance zones are used when geometric control is more important in one direction than another. They apply to a cross-sectional element of a feature, such as straightness applied to a pin's surface.

Straightness is measured in one direction only, lengthwise; not around the pin, which is never intended to be straight.

CHAPTER 10

Machine Terms and Manufacturing Processes

There are some terms used in the machining of parts or other methods of manufacturing that may require clarification. In this chapter we will examine some machining terms, features, and processing instructions that you may not be familiar with. Hopefully the notes that indicate particular features or details will then be easier to understand and have greater meaning when they are seen on a print.

Many terms used when machining parts are usually not indicated directly on the drawing of a part. The current practice is to show the feature needed in a view and specify the dimensions and tolerances necessary to describe the feature. In the past the prints used more notes and language on prints. For example, a title block note may state that all fillets have a .375-inch radius unless otherwise specified. The drawing could show a fillet, but it might not have a direct reference or dimension. The term "fillet" should be known in order to understand the drawing.

Another example is the term "draft." A note may state that all drafts are 10° ± 1°, unless otherwise specified. A drawing may or may not show surfaces with drafts. Unless the term "draft" is understood, the drawing cannot be read properly.

The following are some of the terms that can appear on blueprints and drawings.

- **Chamfer**
To chamfer an edge means to produce a chamfer feature. A chamfer is a removal of a small amount of material providing a slightly angled edge.

• Deburr

A burr is an unwanted sharp edge that extends from a surface. Burrs may be at the end of a shaft, around the edge of a hole, or at the edge of any surface. These burrs are caused by machining operations and, in most cases, must be removed. Deburring is the process of removing unwanted burrs and sharp edges. Burrs and sharp edges may cause injury or interfere with the operation of a part or assembly. Therefore they are usually removed. Deburring may be accomplished by manual or machine methods. Wire brushing, filing, honing, tumbling, or other methods that may be specifically indicated.

It is difficult to produce internal corners that are sharp; they often retain material that deviates from the ideal shape. The print may not specify how sharp internal corners must be, nor how sharp external corners are allowed to be, and simply leave it up to the machinist's common sense.

As described in ISO Standard 13715 the state of either external edges or internal corners can be specified with the "Edge State Symbol". This provides flexibility in how the edges are removed, while also being specific about the distance from the edge that is allowed to be affected by the Deburring process. The production of sharp internal and external corners is not left to chance or to common sense, instead it is specified. The chart below gives several examples.

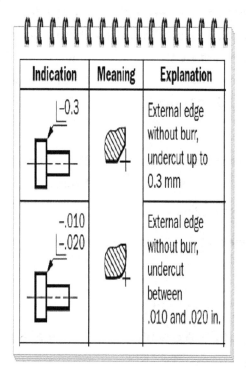
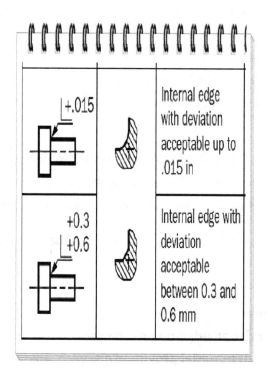

- Knurl

A knurl is a series of small V-shaped grooves machined on a round or internal surface. Knurls are generally diamond-shaped, but they may also be straight. The grooves of a knurl create a rough surface that is used to reduce slippage and is sometimes used for decoration. Knurled surfaces are commonly found on all types of hand tools and instruments. Knurls are machined with hardened knurling wheels and is usually performed on a lathe.

Knurling is typically not shown in detail, but it is specified by type, pitch, and diameter before and/or after knurling. Following are a few examples.

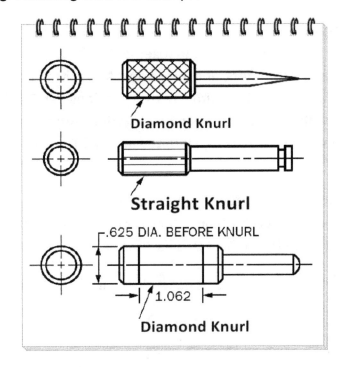

- Serrate

A serrated surface is a flat or curved area that has a series of small V-shaped grooves. Serrations are generally diamond-shaped. They are commonly found on machine tools and are used to reduce slippage.

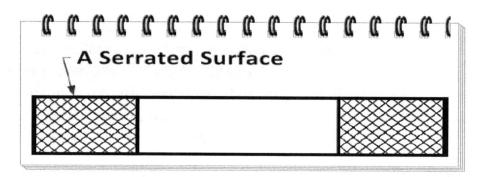

- **Key-seats and Key-ways**

A key is a fastener used to lock a shaft and hub together for the transmission of torque. Keys are typically used with gears, sprockets, and lever arms. A key sits in a key-seat in the shaft that holds the key, and it permits a hub with a key-way to fit over it. The key may be fastened to the shaft, contained within a pocket, or clamped together with a set screw through the hub. Key-seat and key-way geometry is standardized in ASME B17.1 for English keys and DIN 6885 for metric keys.

The prior practice was to specify key-seat and key-way features was with a note. The geometry is standardized, in order to eliminate any confusion and ambiguity, the current practice is to dimension and tolerance all features directly.

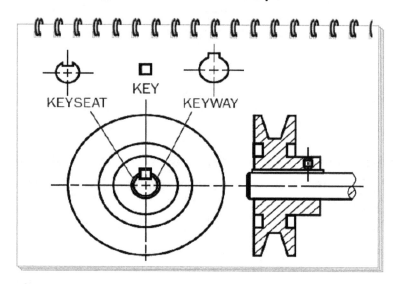

Where features are still found to be indicated by note as was prior practice, the indication KEYSEAT or KEYWAY is shown below.

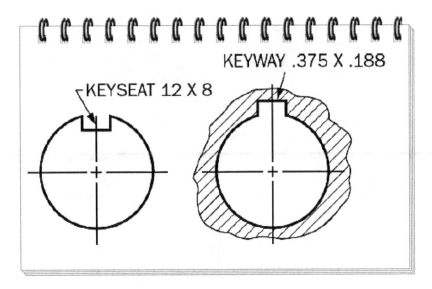

Prior practice rectangular key-seat and keyway indications Key-seat and keyway geometry is shown below. In the English system, the keyway and key-seat are dimensioned by width and depth at the side of the feature. The keyway dimension is indicated by W × T2, and the key-seat dimension is indicated by W × T1. Unless otherwise specified, the shaft keyway is assumed to be standard. A list of standard key and corresponding key-seat and keyway dimensions for English shafts is provided in Table 10-1.

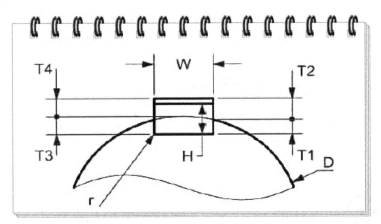

• Neck, Groove and Undercut

A groove cut into a cylindrical surface may be referred to as a "groove," "neck," or "undercut." They are used when a flush fit is necessary; the feature eliminates a radius between the diameter and the shoulder that may interfere with a flush fit.

Prior practice was to indicate these features on prints by note "NECK," "GROOVE," or

"UNDERCUT" and indicating width and depth.

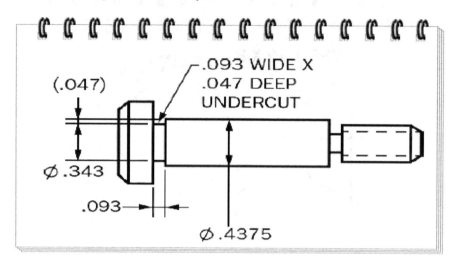

A saw cut feature is a narrow slot through a part produced by a reciprocating, band or circular saw typically for initial roughing out. Saw cuts may be indicated on drawings by a SAWCUT note and width rather than dimensioned directly, indicating that a tight tolerance and good surface finish are not needed.

• Holes

Several round holes produced by various methods are shown below. In prior drawing practice, round holes were commonly indicated by the machining process used to produce them, such as Drill, Ream, Bore, Broach, EDM (e.g., Wire EDM, Plunge EDM), Counter bore, Countersink, Counter drill, Spotface, Drill & Tap, etc.

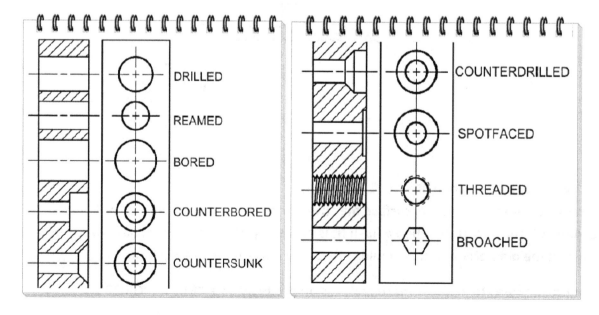

- **DRILL**

The word Drill indicates that a hole to be machined by drilling with a twist drill.

- **REAM**

The word Ream indicates a two-operation process of pre-drilling and then reaming, which removes a small amount of material and achieves a more accurate diameter and smooth finish.

- **BORE**

The word Bore indicates the use of a boring bar to enlarge a rough hole to produce a more accurate diameter and smooth finish. The pre-drill diameter may or may not be indicated for reamed and bored holes. Current practice is to produce prints using symbols that indicate the hole's dimensions and tolerances.

- **Mill**

Milling is one of the most common machining operations, and current drawing practice does not specify the process to be used to produce the feature. In prior practice some prints include terms such as "END MILL" or "FACE MILL," indicating the intended manufacturing practice.

- **Grind**

The word "GRIND" has been used in prior drawing practice to indicate the intended manufacturing practice and expected surface texture.

- **Centers**

Centers, are center drilled holes that are produced on a lathe with a center drill, they are often used to mount the work-piece between the headstock and the tailstock for machining.

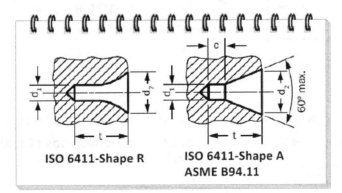

ISO 6411-Shape R

ISO 6411-Shape A
ASME B94.11

The two hole shapes shown above are identified with the letter R or A. The number following the letter indicator is the center size, d1.

Size	d1	c	d2 max	t max
00	.025	.030	1/8	117
0	1/32	.038	1/8	.119
1	3/64	3/64	1/8	.115
2	5/64	5/64	3/16	.175
3	7/64	7/64	1/4	.231
4	1/8	1/8	5/16	.287

Center Drill Sizes

5	3/16	3/16	7/16	.404
6	7/32	7/32	1/2	.462
7	1/4	1/4	5/8	.575
8	5/15	5/16	3/4	.691

Sometimes a general note may appear, giving the machine shop leeway as to the exact size, shape, and even presence of a shaft center.

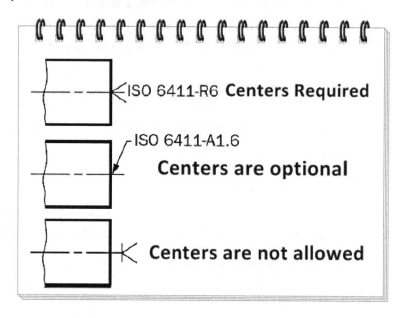

- **Markings**

Parts may be marked with an item number, revision number, serial number, production lot number, company logo, or any other marking. Often used methods of marking parts are shown in the following chart.

ELECTROCHEMICAL ETCH	Uses an electric current to create high-contrast marking without surface distortion
EMBOSS	Uses mechanical stamps, similar to typewriter keys, to indent marking into the surface using hammer or press
INK STAMP	Uses ink that may be applied by a rubber applicator or metal printing die
INKJET	Uses an inkjet marking tool to propel droplets of ink onto the part surface
LASER MARK	Uses a low-powered laser to heat the material, causing oxidation under the surface and turning the material black
LASER ETCH	Uses a laser to melt away the topmost layer of metal, resulting in a raised mark
MARK	Nonspecific term used when the marking technology is not specified
PAINT	Uses paint applied by rubber applicator or stencil
SCRIBE	Uses a sharp stylus to scrape lines including lettering into the part, can be powered or done by hand
DOT PEEN	Uses a sharp stylus to physically indent dots into patterns forming letters on the surface of the part

- **Manufacturing Processes**

Often, there is more than one way to machine a feature of a part. Current practice is for prints to show the finished parts requirements only and not to specify any manufacturing instructions except when following a specific manufacturing process is essential.

In prior practice, general manufacturing instructions were very common even when following the instructions was not mandatory.

- **GENERAL INSTRUCTIONS:**

Some prints use general terms to specify manufacturing operations, and many of these terms have been defined and illustrated in this text. A manufacturer may have a newer, better, or even a different technology available for producing the part, but would be unable to use them because the print specifies particular methods.

A manufacturer interested in using a method contrary to that indicated on the print must request a deviation from the customer, which may also result in a revision of the customer's print.

- **SPECIFIC INSTRUCTIONS:**

Specific manufacturing information may include all or most of the data pertaining to the manufacturing of a part. Specific instructions for machining operations can include equipment, tools, speeds, feeds, coolants etc. Specific instructions for castings can include pattern identifier, sand type, binders, additives, cooling time, descaling methods, and so on. Specific instructions can be found on the print, or they might be found on a process document, which is referenced on the print.

- **FINISHED PART REQUIREMENTS:**

The current practice is for prints is to specify dimensions, but not manufacturing processes. Exceptions to this practice are made if a specific process is essential to meet certain requirements; in this case specific machining instructions are given.

Showing only the finished part requirements allows a manufacturer to choose the method of manufacturing. This practice allows the manufacturer to use the method that best suits their needs and shop equipment. An option of this type will generally provide the best quality at the lowest cost and allow the manufacturer to use newer manufacturing methods that might be better suited to making the part, without requiring print changes.

Blueprints can indicate the finished part requirements only, even though most parts are designed with the manufacturing process that will be used to produce them in mind. A part designed as a forging should contain all necessary information such as, drafted surfaces, rounded corners, and parting lines, etc.

However, an equivalent part designed as a machined part would have square surfaces, sharp corners, and no parting lines. It would be very difficult to make just one print for the manufacture of both a forged part and a machined version of the same part.

- **CHANGING the MANUFACTURING PROCESS:**

It can be relatively easy for a machine shop to substitute one material-removal operation indicated on a print, in favor of another material removal operation.

However, It would be more difficult to manufacture a part as a casting that was originally designed to be machined.

It's unlikely that many parts designed as formed sheet metal would ever be made by machining.

And there are some parts where it is possible to completely change the intended manufacturing process.

Manufacturing technologies are constantly evolving. For example some parts that were designed as castings, have been recently made more economically as weldments. And Parts that were once designed as weldments have been recently made more economically by material-removal processes exclusively, beginning from a large block of metal.

- **HEAT TREATMENT of METALS:**

Many metal parts require heat treating before they can be used. Heat treating is the controlled heating and cooling of a metal to obtain the desired changes in the metals mechanical properties.

Heat treatment is usually indicated by note on the title block. Notes sometimes refer to a specific heat treatment process detailed in related reference documents. Sometimes they can be general notes, or they can leave the methods of heat treating unspecified and indicate only the final hardness required.

It is important to know Some of the terminology related to heat treating, that may appear on your prints.

Note: All metals may be classified as ferrous or nonferrous. A ferrous metal has iron as its main element. A metal is still considered ferrous even if it contains less than 50 percent iron, as long as it contains more iron than any other one metal. A metal is nonferrous if it contains less iron than any other metal.

- Ferrous metals include cast iron, steel, and the various steel alloys, The only difference between iron and steel is the carbon content. Cast iron contains more than 2-percent carbon, while steel contains less than 2 percent.

- Nonferrous metals include a great many metals that are used mainly for metal plating or as alloying elements, such as tin, zinc, silver, and gold. However, this chapter will focus only on the metals used in the manufacture of parts, such as aluminum, magnesium, titanium, nickel, copper, and tin alloys.

- An alloy is a substance composed of two or more elements. Therefore, all steels are an alloy of iron and carbon, but the term "alloy steel" normally refers to a steel that also contains one or more other elements. For example, if the main alloying element is tungsten, the steel is a "tungsten steel" or "tungsten alloy." If there is no alloying material, it is a "carbon steel."

- QUENCH:
The rapid cooling of a heated part in a quenching medium. The quenching medium depends upon the size and shape of the part as well as on the grade of steel. Water, oil, and air are the most common quenching mediums and may be specified. Hardened parts are usually tempered after hardening in order to reduce the hardening stresses and increase the toughness.

- TEMPER:

Tempering is used to lower a metal's brittleness or hardness. It involves heating the steel to below the metal's critical range. The exact temperature depends on the type of steel used and its application.

- **ANNEAL OR NORMALIZE:**

Annealing and normalizing soften the metal to release internal stresses developed in manufacturing processes such as forging. They involve heating the metal above the transformation temperature, holding it there a sufficient time for the structure to transform, and then cooling at a slow rate.

- **STRESS RELIEVE:**

Stress relieving reduces the internal stresses that are the result of mechanical cold working during machining, drawing, bending, or other similar operations. Stresses are relieved by heating the metal to a temperature below its transformation temperature and then allowing it to cool slowly to room temperature.

- **AUSTEMPER-MARTEMPERING:**

Austempering and martempering are treatments given to gray iron to provide heat-treated properties without the high stresses associated with a full quench. Hardness depends on the carbon content and alloy content.

- **SOLUTION TREATMENT:**

Solution treating temperatures vary by alloy, but they usually involve holding the part at a certain temperature and then quenching it in water so that cooling occurs rapidly.

- **AGE HARDENING:**

An age hardening treatment depends on the alloy and can range from no heat (when precipitation takes place at room temperature over a few days) to a longtime process at low temperatures.

When hardening steel, it is heated to a temperature above its transformation temperature and then cooled rapidly. The length of time that it is held at the elevated temperature and the cooling rate depend on the type of steel being treated.

Cast iron can be annealed to soften it, remove residual casting stresses, and improve its machine-ability. Normalizing involves heating the casting and then Cooling it in air; this restores the cast material's properties and improves hardness and tensile strength. Gray iron is hardened by heating it above its transformation temperature and quenching it in oil.

Note: Aluminum alloys cannot be heat-treated in the same way as steel but can be hardened by other processes such as age hardening or solution treating. There are a few scales of hardness that have been developed to measure hardness of materials, each scale has its own testing methodology. Hardness is usually specified using one of the following hardness scales, BRINELL, ROCKWELL, VICKERS.

ROCKWELL HARDNESS TEST

This test determines the hardness of metals by measuring the depth of impression, which can be made by a hard test point under a known load. The softer the metal, the deeper the impression. Soft metals will be indicated by low hardness numbers. Harder metals permit less of an impression to be made, resulting in higher hardness numbers. Rockwell hardness testing is accomplished by using the Rockwell hardness testing machine, shown above.

BRINELL HARDNESS TEST

Brinell hardness testing operates on almost the same principle as the Rockwell test. The difference between the two is that the Rockwell hardness number is determined by the depth of the impression while the Brinell hardness number is determined by the area of the impression. This test forces a hardened ball, 10 mm (0.3937 in) in diameter, into the surface of the metal being tested, under a load of 3,000 kilograms (approximately 6,600 lb). The area of this impression determines the Brinell hardness number of the metal being tested. Softer metals result in larger impressions but have lower hardness numbers.

Perhaps the best known numerical code is the Society of Automotive Engineers (SAE) code. For the metals industry, this organization pioneered in developing a uniform code The SAE system is based on the use of four-or five digit numbers.

- The first number indicates the type of alloy used; for example, 1 indicates a carbon steel. 2 indicates nickel steel.

- The second, and sometimes the third, number gives the amount of the main alloy in whole percentage numbers.

- The last two, and sometimes three, numbers give the carbon content in hundredths of 1 percent (0.01 percent).

Example:

SAE 1045 = 1- Type of steel (carbon).

    .   0- Percent of alloy (none)

    .   45- Carbon content (0.45-percent carbon).

CHAPTER 11

SURFACE FINISH and GEOMETRIC DEVIATIONS:

The quality of the surface finish (The degree of surface smoothness) is important in the manufacture of many parts. The smoothness of a bore is a good example. Normally the smoother the finish of a machined surface, the more expensive it is to manufacture.

The method currently used provides more detailed surface information.

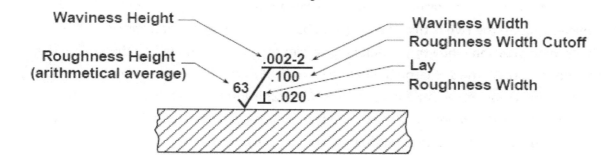

MICRO-INCHES	TYPE OF SURFACE	PURPOSE
1000	EXTRENELY ROUGH	Clearance surfaces where appearance is not important.
500	ROUGH	Used where stress requirements and close tolerances are not required.
250	MEDIUM	Most popular where stress and tolerances requirements

		are essential.
125	AVERAGE SMOOTH	Suitable for mating surfaces and parts held by bolts and rivets with no motion between them.
63	ABOVE AVERAGE FINISH	For close fits or stress parts, except rotating shaft, axles and parts subject to extreme vibration.
32	FINE FINISH	Used where stress concentration is high, and for applications such as bearings.
16	VERY FINE FINISH	Used where smoothness is of primary importance, i.e. high-speed shaft bearings, heavily-loaded bearings and extreme tension members.
8	EXTREMELY FINE FINISH Achieved by grinding, honing, lapping and buffing.	Used for cylindrical surfaces.

- Surface finish refers to the physical texture of a surface produced by machine cutting tools or other fabrication techniques such as molding or casting. The surface finish of a part is created by normal machining processes and sometimes additional mechanical surface processing (such as polishing), heat treatments, or chemical treatments.

- Surface texture or finish refers to the degree of smoothness, or roughness, on any surface of a part. There are many reasons for specifying and controlling surface texture. The way some parts function is primarily determined by their finish. A ball bearing must have smooth surfaces to reduce friction. An automobile must have an engine block with smooth cylinder walls to run efficiently. The handles on a pair of pliers must be rough on the outside, so they will not slip in your hand.

• The surface finish of a machine part can be just as important as its material properties. A surface's geometric and material properties can significantly affect friction, wear, fatigue, corrosion, and electrical and thermal conductivity. Surface texture affects appearance, friction, lubrication, sealing, wear, and many other characteristics.

A surface has a particular shape, roughness, and appearance, but it can also have a layer with properties that differ significantly from those of the bulk material. This is because a machined surface is the result of fracture after an intense shearing process. Measuring and describing surface features and their characteristics are among the most important aspects of machine tool design.

Texture Characteristics:

The next image shows an enlarged section view of a typical machined surface. Many machined surfaces appear to be smooth, but when magnified, they are similar to the image.

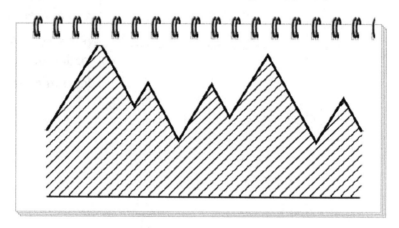

(A view of a typical machined Surface)

The detailed contours of a surface can be split into three main components: roughness, waviness, and form. Additional characteristics of surface texture include lay, skewness, and flaws, can be seen in the next image. For most applications, roughness is the only characteristic that is specified on prints.

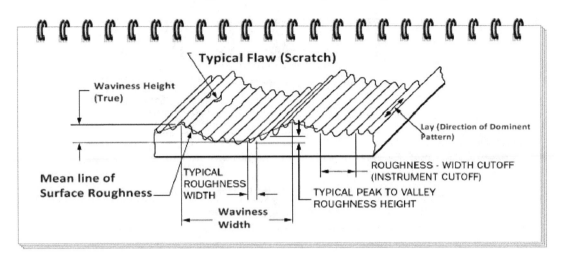

GEOMETRIC DEVIATIONS:

Types of Deviations:

• Surface Discontinuities

• Roughness

• Waviness

• Edge Deviations

• Size Deviations

• Form Deviations

• Orientation Deviations

• Location Deviations

This chart lists the different kinds of geometric deviations. In addition to these requirements the designer also needs to specify the material to be used and any special conditions relating to the material. Complex engineering drawing may require you to understand several different conventions, each controlling different aspects of the design. See the following chart.

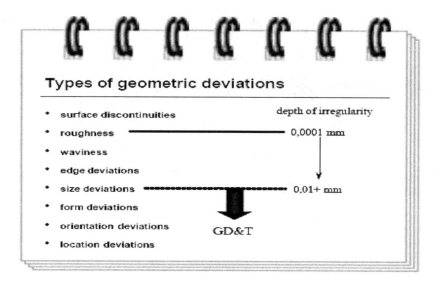

- Only the last three deviations are part of GD&T. Size deviations are controlled by normal dimensioning practices.

- Roughness and waviness require separate specifications.

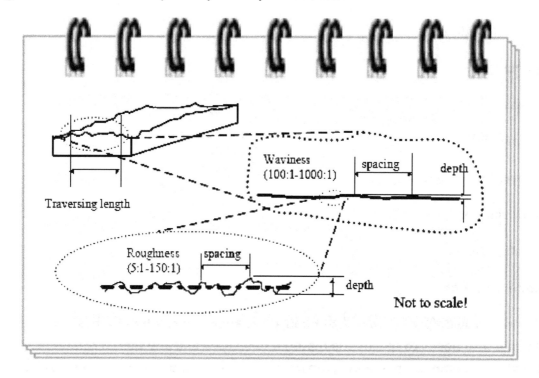

WAVINESS and ROUGHNESS:

ROUGHNESS is a measure at a much smaller scale than waviness and is an indication of very small local imperfections in a surface. Roughness is produced by the direct effect of Machining, Crystallization, Corrosion, or other chemical processes. Marks produced by cutting tools during machining operations are examples of roughness; lines of this type are commonly called "tool marks."

ISO standards define many ways that the roughness profile can be measured and reported. The two most common surface roughness measurements are the arithmetic mean value (Ra) and the root mean square (Rq). Each defines the mean line or mean plane of a surface through different mathematical approaches. Since different calculations are made, the two methods sometimes arrive at different conclusions. It is therefore important not to mix the two. The Ra measurement can be found on most engineering prints; however, it does not capture the full topography of the surface.

Since roughness only describes the average roughness or gives maximum or minimum roughness peak heights, the same roughness measurement can occur for very different surfaces. The figure below shows four surfaces with the same Ra roughness values. Although all these surfaces have the same roughness value, they will each react quite differently in rolling or sliding situations. For critical applications, specifying surface roughness without also defining waviness and skewness is not adequate. Specifying only Ra or Rq is even less informative.

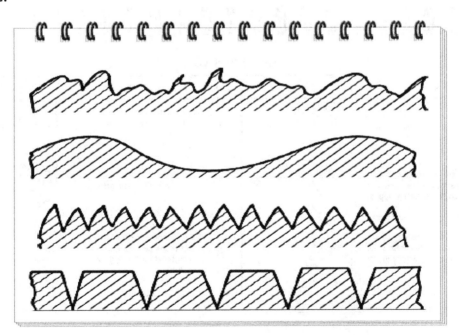

WAVINESS: refers to periodic regularities in the surface of the part, but at a scale smaller than that, which is controlled by GD&T. Waviness, has a longer period than roughness and can be caused by any cyclic characteristic of the machine-tool-tool holder combination. Flexure of the work piece or the tool system is also a cause of waviness. Waviness is at its worst when a critical frequency is reached and chatter occurs.

FORM:

Form errors result from machine characteristics, like variations in slides and incorrect tool settings. Form errors can also occur when thermal distortions of the workpiece, or machine occur during cutting. Improper workpiece support can also cause form errors during machining.

LAY:

All cutting tools leave distinctive marks on the surface of the workpiece that could affect the way the surface wears and or interacts with mating parts. The direction of tool marks depends on the machining method chosen. The general direction may be specified for applications where lay is critical.

LAY SYMBOLS

SYM	DESIGNATION	EXAMPLE	SYM	DESIGNATION	EXAMPLE
—	Lay parallel to the line representing the surface to which the symbol is applied.	DIRECTION OF TOOL MARKS	X	Lay angular in both directions to line representing the surface to which symbol is applied.	DIRECTION OF TOOL MARKS
⊥	Lay perpendicular to the line representing the surface to which the symbol is applied.	DIRECTION OF TOOL MARKS	M	Lay multidirectional	
C	Lay approximately circular relative to the center of the surface to which the symbol is applied.		R	Lay approximately radial relative to the center of the surface to which the symbol is applied.	

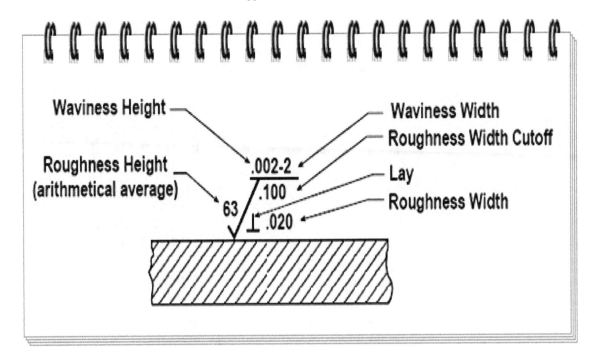

SKEWNESS:

Skewness indicates the degree to which roughness comes from external variation or internal variation. Note: Two surfaces can have the same Ra values but with different skewness. For a contact-bearing application, sharp deep valleys separated by wide flat planes (negative skewness) would perform better than sharp spikes (positive skewness), for the same overall roughness value. A surface with positive skewness would contain spikes that would soon wear off, creating wear debris and damage.

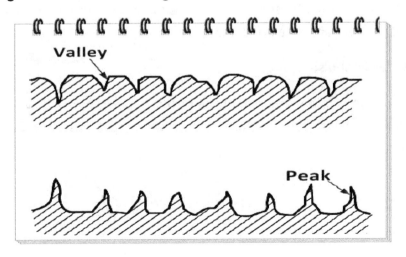

Flaws:

• Flaws are infrequent, random defects such as scratches or cracks.

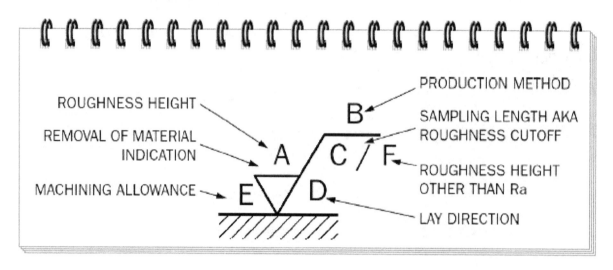

Specification of Surface Texture:

The surface texture symbol looks like a check mark. This symbol provides a uniform and concise way of specifying surface texture. Use of this symbol has been standardized by the American National Standards Institute in ANSI/ASME 46.1 and Y14.36. These standards are recognized and used throughout the world.

The above image presents the basic symbol along with the location of the various surface texture indications for specifying roughness, production method, inspection length, and lay. When requirements other than those listed here are to be designated over a surface (such as skewness or flaws), a separate note is included.

Removal of Material Symbols:

The basic surface texture symbol occurs in three forms: "open", "closed" with a bar, or fitted with a circle to indicate whether the surface, respectively, may, must, or must not be machined, as seen below.

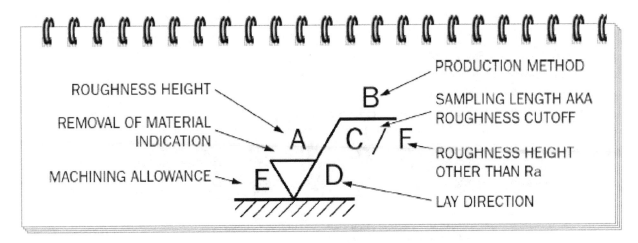

Roughness Height:

The number specifying the roughness height is found in position "A" (see above image). When alone, the number indicates the arithmetic average of the peaks and valleys of the surface, and as such it is referred to as "Ra."

One number above the symbol indicates the maximum roughness. Any surface smoother than the one specified by the symbol is acceptable.

Surface texture for English prints is specified in micro inches [millionths (.000001) of an inch]. Surface texture for metric prints is specified in micrometers [thousandths (0.001) of a millimeter]. Table 8-1 indicates standard roughness values for English and metric prints. The values indicated are approximately equivalent (63 micro inches is approximately 1.6 micrometers).

Specifying only one maximum number on a print may lead to producing a surface that is smoother than necessary at an added expense. Producing a surface that is too smooth may also be a physical problem for some applications.

Two numbers, as seen below, indicate a minimum and maximum roughness value. If a different measurement technique from Ra is intended, the number must be preceded by the appropriate symbol, such as "Rt" or "Rz," and the symbol and number are found in position "F".

SIZE DEVIATION:

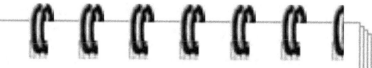

- deviation of actual local size from nominal linear size or from nominal angular size
- actual local linear sizes are assessed by 2-point measurements
- actual local angular sizes are assessed by angular measurements of averaged lines
- assessed over entire geometric element
- produced mainly by imprecise adjustment of machine tool and cutting or process conditions

Size deviation is controlled on engineering drawings by starting the nominal size and a tolerance, which defines the maximum permissible deviation from the nominal size. The nominal size and tolerance define the permissible design limits both upper and lower, within all size measurements must lie.

When the tolerance values $\pm$ on either side of the nominal have the same values it is known as "Bilateral Tolerancing."

However, when the values are different, we call it "Unilateral Tolerancing." Unilateral Tolerancing is used by designers to influence the machinist to opt for one tolerance $\pm$ over another, to boost the efficiency and or production rate of a part. While manufacturing designs are created using Unilateral Tolerancing, it is considered "Normal practice" to convert Unilateral to Bilateral Tolerances during the planning stages.

FORM DEVIATION:

- Deviation of a feature (geometrical element, surface or line from its nominal form.

- The deviation is assessed over the entire feature, unless it is otherwise specified.

- If the spacing to depth ratio is greater than 1000:1

Form Deviations are produced by:
- Problems in the guide-ways and or bearings of the machine.
- The DEFLECTION of the cutting tool and or fixture.
- Problems with the fixture
- Tool and or fixture WEAR.

Form deviation is the deviation of a feature nominal (Dimension) from its shape. This feature could be a line on a surface, the surface itself, or a geometric element. An axis would be one such example. A form deviation is specified without reference to other features.

All geometric deviations are assumed to apply over the entire feature unless otherwise specified. A straightness control applied to lines on a surface means the entire surface is to be controlled, not just a portion of it.

Geometric Characteristic Symbols				Modifying Symbols	
	Type of Tolerance	Characteristic	Symbol	Term	Symbol
For individual features	Form	Straightness	—	At maximum material condition	Ⓜ
		Flatness	▱	At least material condition	Ⓛ
		Circularity (roundness)	○	Projected tolerance zone	Ⓟ
		Cylindricity	⌭	Free state	Ⓕ
For individual or related features	Profile	Profile of a line	⌒	Tangent plane	Ⓣ
		Profile of a surface	⌓	Diameter	⌀
For related features	Orientation	Angularity	∠	Spherical diameter	S⌀
		Perpendicularity	⊥	Radius	R
		Parallelism	//	Spherical radius	SR
	Location	Position	⊕	Controlled radius	CR
		Concentricity	◎	Reference	( )
	Runout	Symmetry	≡	Arc length	⌒
		Circular runout *	↗	Statistical tolerance	⟨ST⟩
		Total runout *	⤴	Between *	↔

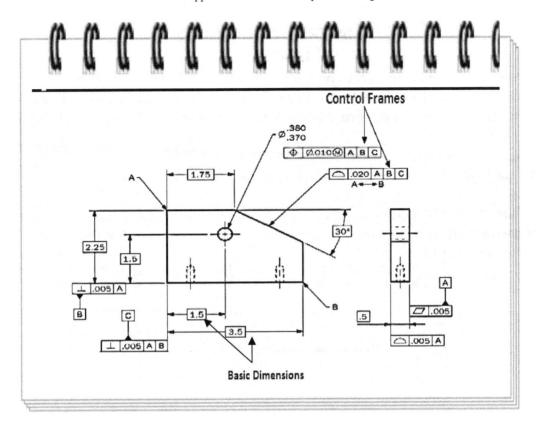

1. The geometric characteristic symbol: This symbol indicates what type of geometric tolerance applies . In this example, the first box tells that we are dealing with perpendicularity.
2. Next, you have the shape of the tolerance zone, (Optional): The shape will be identified by one of the following: a square symbol, a diameter symbol, or a spherical diameter (rarely), or it can be left blank. When no symbol is given, the shape of the tolerance zone is inferred from the shape of the feature.
3. The total size of the tolerance zone: This is always given in the same linear dimension units used on the rest of the print—inches or millimeters. In the image above, we can see the size of our tolerance zone, .008.
4. The material modifier symbol (optional): The symbol will be one of these three: the MMC symbol, the LMC symbol, or the S symbol or no symbol at all. For MMC, bonus tolerance is available on an MMC basis; and for LMC, bonus tolerance is available on an LMC basis. If no symbol is given or if the S symbol is used, there is no bonus tolerance allowed.

A fundamental rule of geometric dimensioning and tolerancing is that a size dimension controls not only the size of the feature, but also its form. Any feature with size that is produced at its MMC is only acceptable if its form is perfect.

MATERIAL CONDITIONS:
Modifiers:

GD&T allows for certain modifiers to be used in specifying tolerances at various part feature conditions. These conditions may be the largest size of the feature, the smallest size of the feature or actual size of that feature. Material conditions may only be used when referring to a feature of size.

Material conditions can only be used when referring to a feature of size.

A Maximum Material Condition occurs when a feature of size contains the maximum amount of material everywhere within the stated limits of size. This could be the largest shaft diameter or smallest hole.

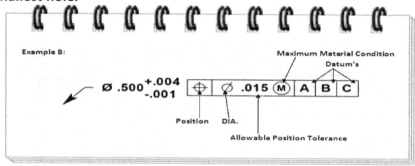

Remember:

• The Maximum Material Condition (MMC) of an external feature of size, is the features largest size limit.

• The Maximum Material Condition (MMC) of an internal feature of size, is the smallest size limit.

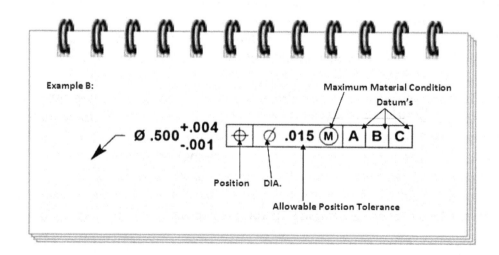

Example B:
- If a hole, for instance, has the following size and geometric control, and the hole measures .502.
- It would be incorrect to use a bonus tolerance of .003 (.502 - .499(MMC)) if the hole is not perfectly oriented to the Datum's.
- If the hole is out of perpendicular to datum A by .002, for instance, the bonus that may be used is reduced by that amount. The bonus would be merely .001 and the allowable position tolerance = .016.

Geometric Characteristic Symbols				Modifying Symbols	
	Type of Tolerance	Characteristic	Symbol	Term	Symbol
For individual features	Form	Straightness	—	At maximum material condition	Ⓜ
		Flatness	▱	At least material condition	Ⓛ
		Circularity (roundness)	○	Projected tolerance zone	Ⓟ
		Cylindricity	⌭	Free state	Ⓕ
For individual or related features	Profile	Profile of a line	⌒	Tangent plane	Ⓣ
		Profile of a surface	⌓	Diameter	⌀
For related features	Orientation	Angularity	∠	Spherical diameter	S⌀
		Perpendicularity	⊥	Radius	R
		Parallelism	//	Spherical radius	SR
	Location	Position	⌖	Controlled radius	CR
		Concentricity	◎	Reference	( )
		Symmetry	⌯	Arc length	⌒
	Runout	Circular runout	↗	Statistical tolerance	Ⓢ⃝Ⓣ
		Total runout	↗↗	Between	↔

The rectangular box is called a feature control frame and indicates a geometric tolerance. The proper way to read a feature control frame is shown below:

The proper way to read a feature control frame, is from left to right. As you can see, a feature control frame consists of a series of symbols, letters and numbers. Variation from perfect form is only allowed to the extent that the size varies from MMC. Some form variation is allowed (bow, necking, barreling, out-of-round, etc.), but only variation that does not push the material of the pin beyond its 16-millimeter

Limits of Accuracy:

All machinists must work within the limits of accuracy, as specified on the blueprint. As a machinist, understanding tolerance and allowance will help you in making good parts. These terms, "Tolerance" and "Accuracy" may seem to be related, but each has a very different meaning and application. In the following pages, you will learn the meanings of these terms and the difference between them.

Bonus Tolerance:

In some cases, the geometric tolerance given in a feature control frame can actually be made larger during inspection, allowing more variation than the print would seem to indicate. This tolerance is sometimes referred to as "bonus tolerance," because it is added to the geometric tolerance in the feature control frame to get the total tolerance used to verify the feature. Bonus tolerance maybe indicated on the print for a size feature, but never for a surface. Bonus tolerance is indicated in one of two ways, on a maximum material condition (MMC) basis or on a least material condition (LMC) basis. The basis for the bonus tolerance is indicated by the symbol in the feature control frame after the geometric tolerance. With the MMC modifier, the geometric tolerance given on the drawing applies when the size of the feature actually produced is its maximum material condition; for a feature produced at any other size, there is bonus tolerance. The RFS symbol after the geometric tolerance, or no symbol, indicates there is no bonus tolerance for the feature.

The calculation involves only two numbers: the actual produced size of the feature and either its MMC or LMC. The MMC or LMC can be calculated from the dimension and tolerance on the print; the actual produced size of the feature is measured at inspection. Only one calculation is used, depending on whether the feature is internal or external and on what modifier is used; see table below.

Calculating Bonus Tolerance and Total Tolerance for Features

	Internal Feature	External Feature
MMC	Actual produced Size - Maximum material condition = Bonus Tolerance.	Maximum material condition – Actual produced size = Bonus Tolerance.
LMC	Least material condition – Actual produced size = Bonus Tolerance.	Actual produced size – Least material condition = Bonus Tolerance.

Total Tolerance = Tolerance from feature Control Frame + Bonus
LIMIT TOLERANCE:

When a dimension has a high and low limit stated. 1.250/1.150 is a limit tolerance.

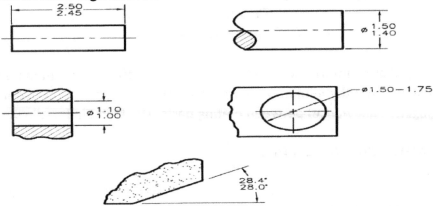

PLUS or MINUS TOLERANCE:
The nominal or target value of the dimension is given first, followed by a plus-minus expression of tolerance. .32 +/- .004 is a plus-minus tolerance.

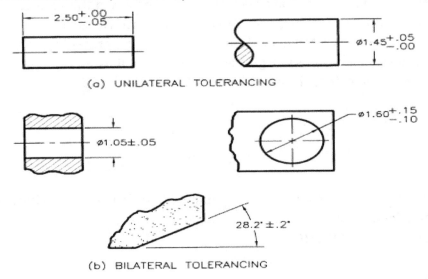

BILATERAL TOLERANCE:
A tolerance that allows the dimension to vary in both the plus and minus directions.

EQUAL BILATERAL TOLERANCE:
Variation from the nominal is the same in both directions
UNILATERAL TOLERANCE:
Where allowable variation is only in one direction and zero in the other.
UNEQUAL BILATERAL TOLERANCE:
Where the allowable variation is from the target value and the variation is not the same in both directions.

TYPES OF FIT:
ALLOWANCE

Allowance is defined as an international difference between the maximum material limits of mating parts. Allowance is the minimum clearance (positive allowance), or maximum interference (negative allowance) between mating parts. The formula for allowance is:

ALLOWANCE = MMC HOLE – MMC SHAFT

Allowance is an intentional difference in dimensions of mating parts to provide the desired fit. A clearance allowance permits movement between mating parts when assembled. For example, when a hole with a 0.250-inch diameter is fitted with a shaft that has a 0.245-inch diameter, the clearance allowance is 0.005 inch. An interference allowance is the opposite of a clearance allowance. The difference in dimensions in this case provides a tight fit. Force is required when assembling parts, which have an interference allowance. If a shaft with a 0.251- inch diameter is fitted in the hole identified in the preceding example, the difference between the dimensions will give an interference allowance of 0.001 inch. As the shaft is larger than the hole, force is necessary to assemble the parts.

CLEARANCE

Clearance is defined as the loosest fit or the maximum intended difference between mating parts. The formula for clearance is:
CLEARANCE = LMC HOLE – LMC SHAFT

- Clearance of FIT

The parts are toleranced in such a way that the largest shaft is smaller than the smallest hole. The allowance is positive and greater than zero.

- Interference Fit

The maximum clearance is always negative.
The parts are always forced together.

- Transition Fit

The parts are toleranced so that the allowance is negative and the maximum clearance is positive.
The parts can be loose or can be forced together.

GEOMETRIC DEVIATIONS:

Types of Deviations:
- Surface Discontinuities
- Roughness
- Waviness
- Edge Deviations
- Size Deviations
- Form Deviations
- Orientation Deviations
- Location Deviations

This chart lists the different kinds of geometric deviations. In addition to these requirements the designer also needs to specify the material to be used and any special conditions relating to the material. Complex engineering drawing may require you to understand several different conventions, each controlling different aspects of the design. See the following chart.

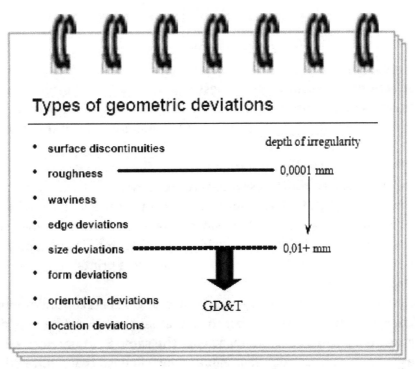

- Only the last three deviations are part of GD&T. Size deviations are controlled by normal dimensioning practices.
- Roughness and waviness require separate specifications.

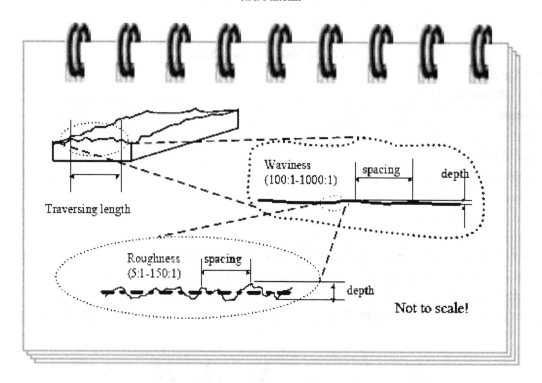

WAVINESS and ROUGHNESS:

ROUGHNESS is a measure at a much smaller scale than waviness and is an indication of very small local imperfections in a surface. Roughness is produced by the direct effect of Machining, Crystallization, Corrosion, or other chemical processes. Marks produced by cutting tools during machining operations are examples of roughness; lines of this type are commonly called "tool marks."

ISO standards define many ways that the roughness profile can be measured and reported. The two most common surface roughness measurements are the arithmetic mean value (Ra) and the root mean square (Rq). Each defines the mean line or mean plane of a surface through different mathematical approaches. Since different calculations are made, the two methods sometimes arrive at different conclusions. It is therefore important not to mix the two. The Ra measurement can be found on most engineering prints; however, it does not capture the full topography of the surface.

Since roughness only describes the average roughness or gives maximum or minimum roughness peak heights, the same roughness measurement can occur for very different surfaces. The figure below shows four surfaces with the same Ra roughness values. Although all these surfaces have the same roughness value, they will each react quite differently in rolling or sliding situations. For critical applications, specifying surface roughness without also defining waviness and skewness is not adequate. Specifying only Ra or Rq is even less informative.

WAVINESS refers to periodic regularities in the surface of the part, but at a scale smaller than that, which is controlled by GD&T. Waviness, has a longer period than roughness and can be caused by any cyclic characteristic of the machine-tool-tool holder combination. Flexure of the work piece or the tool system is also a cause of waviness. Waviness is at its worst when a critical frequency is reached and chatter occurs.

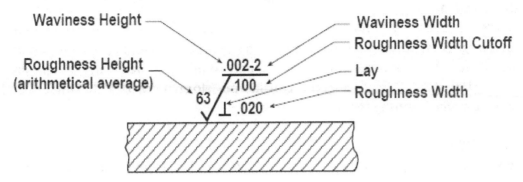

SKEWNESS:
Skewness indicates the degree to which roughness comes from external variation or internal variation. Note: Two surfaces can have the same Ra values but with different skewness. For a contact-bearing application, sharp deep valleys separated by wide flat planes (negative skewness) would perform better than sharp spikes (positive skewness), for the same overall roughness value. A surface with positive skewness would contain spikes that would soon wear off, creating wear debris and damage.

SIZE DEVIATION:

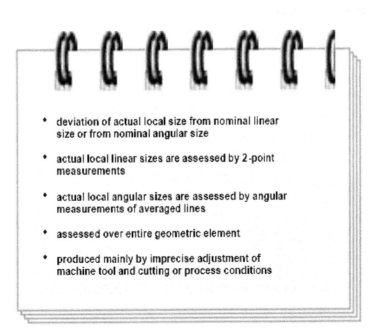

Size deviation is controlled on engineering drawings by starting the nominal size and a tolerance, which, defines the maximum permissible deviation from the nominal size. The nominal size and tolerance define the permissible design limits both upper and lower, within all size measurements must lie.

When the tolerance values ± on either side of the nominal have the same values it is known as "Bilateral Tolerancing."

But when the values are different, we call it "Unilateral Tolerancing." Unilateral Tolerancing is used by designers to influence the machinist to opt for one tolerance ± over another, to boost the efficiency and or production rate of a part. While manufacturing designs are created using Unilateral Tolerancing, it is considered "Normal practice" to convert Unilateral to Bilateral Tolerances during the planning stages.

FORM DEVIATION:
- Deviation of a feature (geometrical element, surface or line from its nominal form.
- The deviation is assessed over the entire feature, unless it is otherwise specified.
- If the spacing to depth ratio is greater than 1000:1

Form Deviations are produced by:
- Problems in the guide-ways and or bearings of the machine.
- The DEFLECTION of the cutting tool and or fixture.
- Problems with the fixture
- Tool and or fixture WEAR.

Form deviation is the deviation of a feature nominal (Dimension) from its shape. This feature could be a line on a surface, the surface itself, or a geometric element. An axis would be one such example. A form deviation is

CHAPTER 12
• Inspection Techniques •

First we will take a look at some common inspection tools.

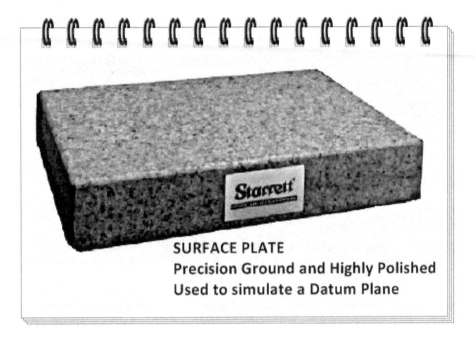

SURFACE PLATE
Precision Ground and Highly Polished
Used to simulate a Datum Plane

Granite surface plates provide a reference surface plane from which final dimensions can be taken. This reference plane is used is used for work inspection and precision layout.

INSPECTING ANGLES:

ANGULARITY:

The Angularity measurement is also a Total Indicator Reading (T.I.R.) The flatness is also inherently controlled.

When checking Angularity a Sine Bar, Sine Plate or CMM Could be used.

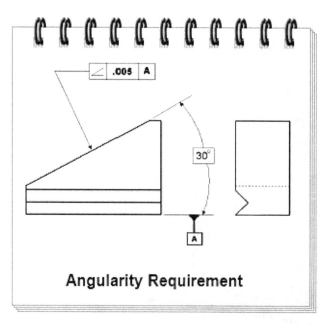

Angularity Requirement

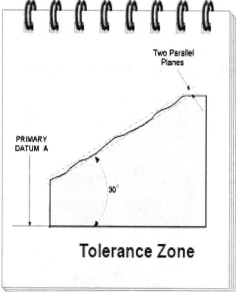

Tolerance Zone

The next image shows a workpiece on which an angle of 30° is required.
• The workpiece is rested on a parallel, which has been wrung to angle blocks the 30° angle.

• Note: the entire set-up is lined up vertically with an angle plate.

• The indicator is swept over the top of the workpiece to determine if the angle is correct.

(Note: It is not necessary to use parallels, however at times they can be useful because of their longer reference surface.)

Inspecting a Simple Angle

Pictured in the above image is a dial indicator, mounted on a

Below are some of the measurement methods used to check compliance to geometric tolerances.

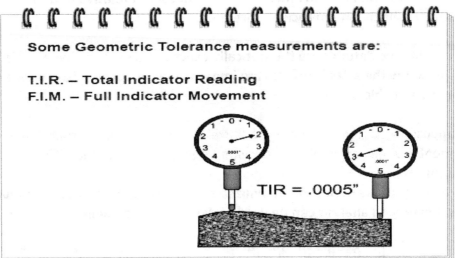

The workpiece must be correctly aligned in the measurement device, otherwise large errors in measurement can occur.

This is defined by the minimum alignment requirements:

• For form tolerances

• For orientation, location and run-out tolerances

• For roundness and cylindricity this requires that the concentric circles and coaxial cylinders be located in such a manner so that their radial distance is at a minimum.

• For straightness and flatness the distance between parallel lines or planes must be kept at a minimum.

In addition to the above requirements, roundness and cylindricity require that the axis of the workpiece not be inclined relative to the measurement device. Any inclination will result in an oval shape and increase the estimate of the amount of deviation.

The Minimum Rock (movement) Requirement:

The minimum rock requirement positions the workpiece relative to the datum feature. Such as when the amount of rock (movement) in all directions is equalized.

Deviations for orientation, location and run-out are measured relative to datum's. The datum's must meet the minimum rock requirement.

On drawing callouts the datum's are theoretically exact surfaces. They have perfect form and position. As you know the actual surfaces are imperfect, a supposedly flat surface may in reality be convex and able to rock.

Likewise a supposedly straight shaft may be barrel shaped and able to rock. The minimum rock requirement is aimed at providing consistent alignment under the different conditions that might occur.
Note: When rocking occurs it must be minimized by equalizing the amount of movement in all directions. Regression analysis can be used to determine the minimum rock orientation.

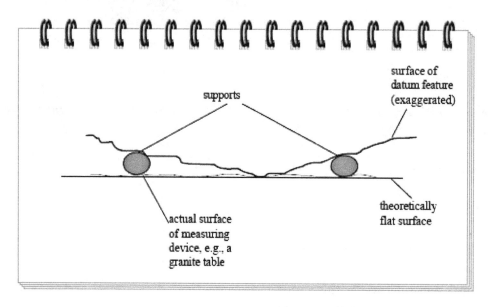

The above image shows the minimum rock requirement for a datum feature. The datum feature will be positioned on the measuring device, which should have a surface condition of at least an order of magnitude better than the datum feature. Supports can be used to minimize the amount of rock of the datum feature, as shown in the above image.

Some geometric tolerances can result in costly inspection methods, so simplified inspection methods should be used at first. If the deviations exceed the tolerance zone then the actual inspection method is used, as an example run-out is easy to determine and not very costly. The method used for checking run-out can be used as a simplified method for checking concentricity and straightness of axis, which are more costly to determine.

Precise determination of the minimum rock requirement is usually very costly. So approximate inspection methods can be used, with the aid of v-blocks, mandrels, center holes, etc.

Approximate methods will result in a greater estimate of deviation, than actually exist, making them more conservative. So while bad parts are rejected, some marginally good parts will also be rejected. This practice is considered as acceptable practice by some because of the reduction in inspection costs.

FLATNESS:

Flatness is a tolerance on a flat part, when it is checked correctly, it will control the flatness and straightness of the surface of the part and the size as well.

The plate in the image below must fit through an envelope not greater than 3+0.05+0.1=3.15 mm above the plate, the straightness can be verified. See image below.

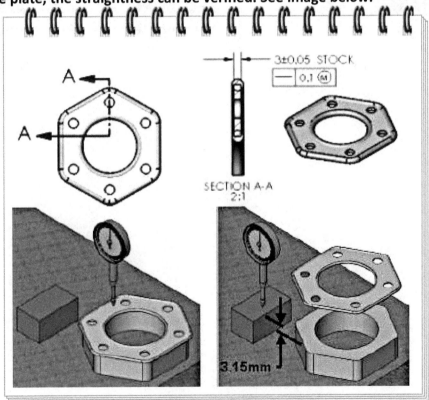

CHECKING FLATNESS TOLERANCE:

The surface must be leveled (at its optimum plane) so that the indicator or probe reads only the hills and valleys of the flatness error.

Jack Screws Method

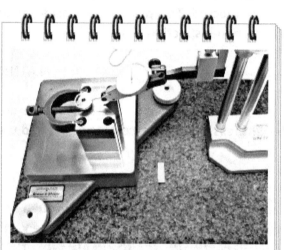

Leveling ("Wobble") Plate Method

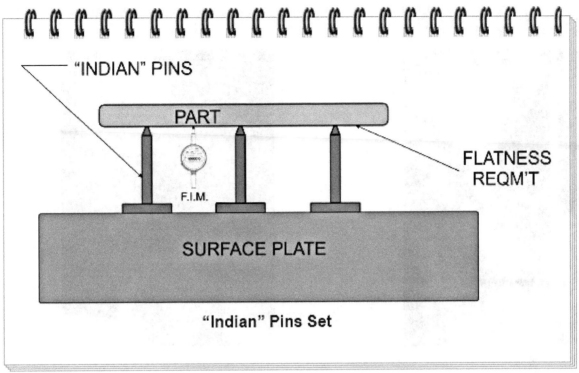

STRAIGHTNESS OF SURFACE ELEMENTS:

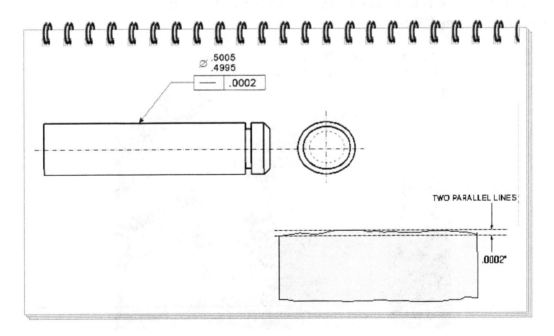

A typical set-up with two Jack Screws, a V-Block and a Parallel. The jacks are used to level the line element and reach a top dead center total indicator reading (TIR).

 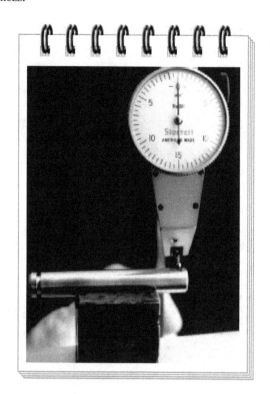

An alternative method is to build two equal gage block stacks, then a bottom dead center total indicator reading (TIR).

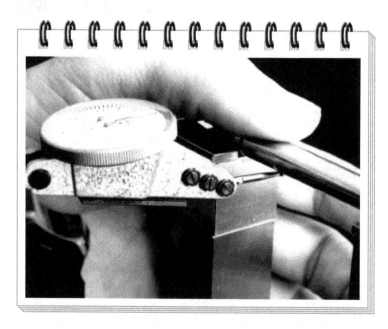

STRAIGHTNESS OF AXIS RFS:

Differential measurements are required for inspection. Two opposing indicators are used to track axial deviation.

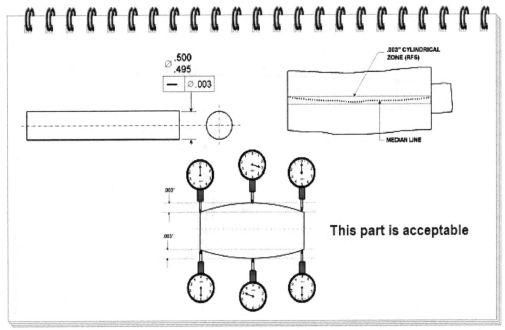

CIRCULARITY (ROUNDNESS):

A roundness tolerance zone is made up of two concentric circles. Note. Roundness is a radial measurement, not a diametral measurement.

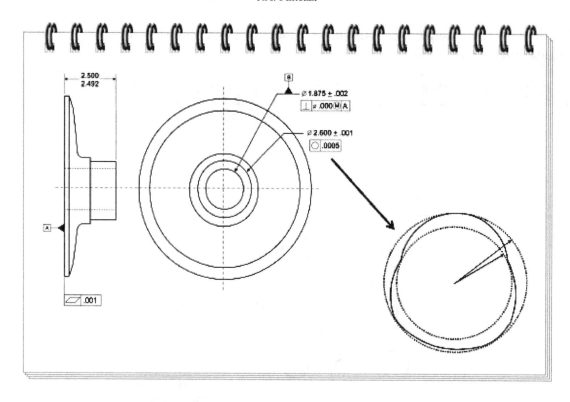

When checking circularity a Precision Spindle or CMM could be used.

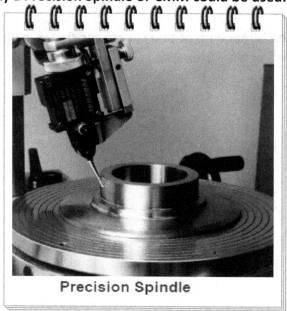

PARALLELISM OF A SURFACE:

The Parallelism is a TIR once the datum has been mounted.

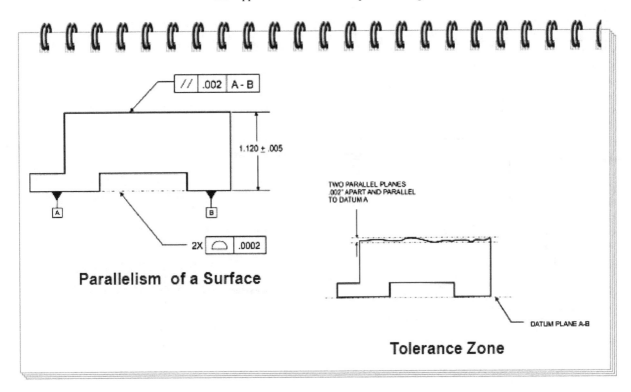

The dial indicator is swept across the entire surface. The resulting TIR should not exceed the tolerance. The flatness is automatically controlled.

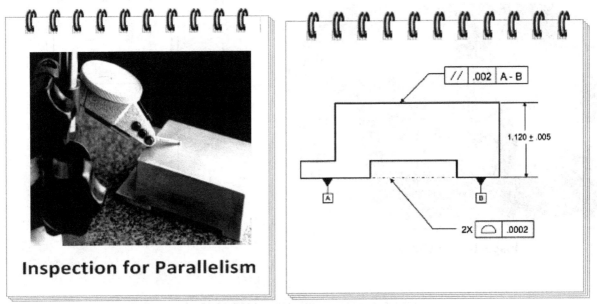

PERPENDICULARITY OF A SURFACE:

The perpendicularity measurement is also a T.I.R. The flatness is inherently controlled.

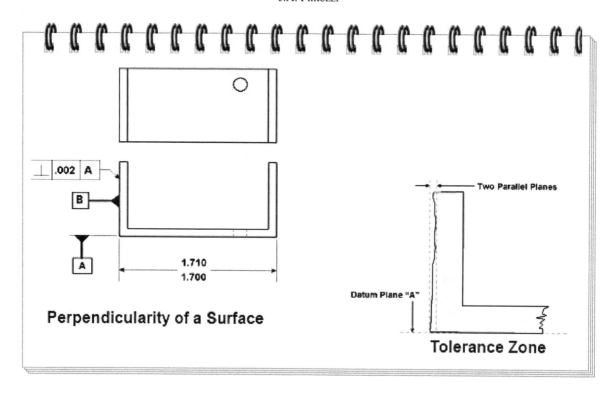

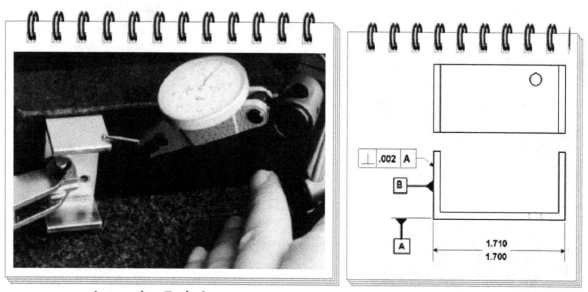

PERPENDICULARITY OF A SURFACE:
(SECOND DATUN)

The part is aligned like this when there is a secondary datum.

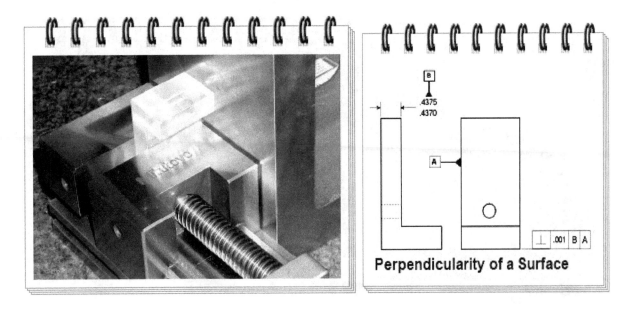

ANGULARITY:

The Angularity measurement is also a T.I.R. The flatness is also inherently controlled.

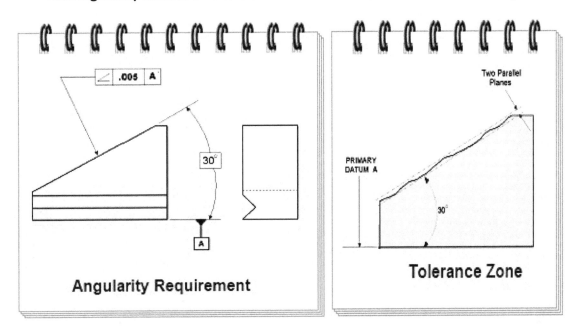

When checking Angularity a Sine Bar, Sine Plate or CMM Could be used.

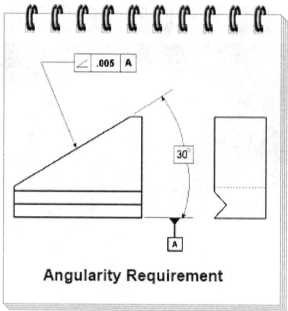

Angularity Requirement

Further Reading..........

- **Interpolation of Geometric Dimensioning & Tolerancing   by Daniel E. Puncochar**

- **Geometric Dimensioning and Tolerancing   by Alex Krulikowski**

- **Geo-Metrics III: The Application of Geometric Dimensioning and Tolerancing Techniques (Using Customary Inch Systems)   by Lowell W. Foster**

- **Tolerance design: a handbook for developing optimal specifications   by C.M. Creveling**
- **Dimensioning and Tolerancing Handbook   by Paul J. Drake**

- **Inspection and Gaging   by Clifford W. Kennedy**

- **Geometric Dimensioning and Tolerancing   by Cecil H. Jensen**

- **ANSI American National Standard Institute**

The Apprentices Guide to Blueprint Reading

GLOSSARY

When you begin a new career in the machine tool trades, you will need to learn the vocabulary of the trade in order to understand your co-workers and to make yourself understood by them. There are many reasons, but most of them boil down to convenience and safety. Under certain circumstances, one word or several words may mean the exact same thing or a certain sequence of actions, making it unnecessary to give a lot of explanatory details. A great deal of the work a machinist does, is such that an incorrectly interpreted instruction could cause confusion, breakage of machinery, or even loss of life. Avoid this confusion and possible danger by learning the meaning of terms common to manufacturing. This glossary is not all-inclusive, but it does contain many terms that every craftsman should know. The terms given in this glossary may have more than one definition; only those definitions as related to the machinist are given.

ALIGNED SECTION—A section view in which some internal features are revolved into or out of the plane of the view.

ANALOG—The processing of data by continuously variable values.

ANGLE—A figure formed by two lines or planes extending from, or diverging at, the same point.

APPLICATION BLOCK—A part of a drawing of a subassembly showing the reference number for the drawing of the assembly or adjacent subassembly.

ARC—A portion of the circumference of a circle.

ARCHITECT'S SCALE—The scale used when dimensions or measurements are to be expressed in feet and inches.

AUXILIARY VIEW—An additional plane of an object, drawn as if viewed from a different location. It is used to show features not visible in the normal projections.

AXIS—The centerline running lengthwise through a screw.

AXONOMETRIC PROJECTION—A set of three or more views in which the object appears to be rotated at an angle, so that more than one side is seen.

BEND ALLOWANCE—An additional amount of metal used in a bend in metal fabrication.

BILL OF MATERIAL—A list of standard parts or raw materials needed to fabricate an item.

BISECT—To divide into two equal parts.

BLOCK DIAGRAM—A diagram in which the major components of a piece of equipment or a system are represented by squares, rectangles, or other geometric figures, and the normal order of progression of a signal or current flow is represented by lines.

BLUEPRINTS—Copies of mechanical or other types of technical drawings. Although blueprints used to be blue, modem reproduction techniques now permit printing of black-on-white as well as colors.

BODY PLAN—An end view of a ship's hull, composed of superimposed frame lines.

BORDER LINES—Dark lines defining the inside edge of the margin on a drawing.

BREAK LINES—Lines to reduce the graphic size of an object, generally to conserve paper space. There are two types: the long, thin ruled line with freehand zigzag and the short, thick wavy freehand line.

BROKEN OUT SECTION—Similar to a half section; used when a partial view of an internal feature is sufficient.

BUTTOCK LINE—The outline of a vertical, longitudinal section of a ship's hull.

CABINET DRAWING—A type of oblique drawing in which the angled receding lines are drawn to one-half scale.

CANTILEVER—A horizontal structural member supported only by one end.

CASTING—A metal object made by pouring melted metal into a mold

CAVALIER DRAWING—A form of oblique drawing in which the receding sides are drawn full scale, but at 45° to the orthographic front view.

CENTER LINES—Lines that indicate the center of a circle, arc, or any symmetrical object; consist of alternate long and short dashes evenly spaced.

CIRCLE—A plane-closed figure having every point on its circumference (perimeter) equidistant from its center.

CIRCUMFERENCE—The length of a line that forms a circle.

CLEVIS—An open-throated fitting for the end of a rod or shaft, having the ends drilled for a bolt or a pin. It provides a hinging effect for flexibility in one plane.

COLUMN—High-strength vertical structural members.

COMPUTER-AIDED DRAFTING (CAD)—A method by which engineering drawings may be developed on a computer.

COMPUTER-AIDED MANUFACTURING (CAM)—A method by which a computer uses a design to guide a machine that produces parts.

COMPUTER LOGIC—The electrical processes used by a computer to perform calculations and other functions.

CONE—A solid figure that tapers uniformly from a circular base to a point.

CONSTRUCTION LINES—Lightly drawn lines used in the preliminary layout of a drawing.

CORNICE—The projecting or overhanging structural section of a roof.

CREST—The surface of the thread corresponding to the major diameter of an external thread and the minor diameter of an internal thread

CUBE—Rectangular solid figure in which all six faces are square.

CUTTING PLANE LINE—A line showing where a theoretical cut has been made to produce a section view.

CYLINDER—A solid figure with two equal circular bases.

DEPTH—The distance from the root of a thread to the crest, measured perpendicularly to the axis.

DESIGNER'S WATERLINE—The intended position of the water surface against the hull.

DEVELOPMENT—The process of making a pattern from the dimensions of a drawing. Used to fabricate sheet metal objects.

DIGITAL—The processing of data by numerical or discrete units.

DIMENSION LINE—A thin unbroken line (except in the case of structural drafting) with each end terminating with an arrowhead; used to define the dimensions of an object. Dimensions are placed above the line, except in structural drawing where the line is broken and the dimension placed in the break

DRAWING NUMBER—An identifying number assigned to a drawing or a series of drawings.

DRAWINGS—The original graphic design from which a blueprint may be made; also called plans.

ELECTROMECHANICAL DRAWING—A special type of drawing combining electrical symbols and mechanical drawing to show the position of equipment that combines electrical and mechanical features.

ELEMENTARY WIRING DIAGRAM— A schematic diagram; the term *elementary wiring diagram* is sometimes used interchangeably with schematic diagram, especially a simplified schematic diagram.

ELEVATION—A four-view drawing of a structure showing front, sides, and rear.

ENGINEER'S SCALE—The scale used whenever dimensions are in feet and decimal parts of a foot, or when the scale ratio is a multiple of 10.

EXPLODED VIEW—A pictorial view of a device in a state of disassembly, showing the appearance and interrelationship of parts.

EXTERNAL THREAD—A thread on the outside of a member. Example: a thread of a bolt.

FALSEWORK—Temporary supports of timber or steel sometimes required in the erection of difficult or important structures.

FILLET—A concave internal corner in a metal component, usually a casting.

FINISH MARKS—Marks used to indicate the degree of smoothness of finish to be achieved on surfaces to be machined

FOOTINGS—Weight-bearing concrete construction elements poured in place in the earth to support a structure.

FORGING—The process of shaping heated metal by hammering or other impact.

FORMAT—The general makeup or style of a drawing.

FRAME LINES—The outline of transverse plane sections of a hull.

FRENCH CURVE—An instrument used to draw smooth irregular curves.

FULL SECTION—A sectional view that passes entirely through the object.

HALF SECTION—A combination of an orthographic projection and a section view to show two halves of a symmetrical object.

HATCHING—The lines that are drawn on the internal surface of sectional views. Used to define the kind or type of material of which the sectioned surface consists.

HELIX—The curve formed on any cylinder by a straight line in a plane that is wrapped around the cylinder with a forward progression.

HIDDEN LINES—Thick, short, dashed lines indicating the hidden features of an object being drawn.

INSCRIBED FIGURE—A figure that is completely enclosed by another figure.

INTERCONNECTION DIAGRAM—A diagram showing the cabling between electronic units, as well as how the terminals are connected

INTERNAL THREAD—A thread on the inside of a member. Example: the thread inside a nut.

ISOMETRIC DRAWING—A type of pictorial drawing. *See* ISOMETRIC PROJECTION.

ISOMETRIC PROJECTION—A set of three or more views of an object that appears rotated, giving the appearance of viewing the object from one corner. All lines are shown in their true length, but not all right angles are shown as such.

ISOMETRIC WIRING DIAGRAM—A diagram showing the outline of a ship, an aircraft, or other structure, and the location of equipment such as panels and connection boxes and cable runs.

JOIST—A horizontal beam used to support a ceiling.

KEY—A small wedge or rectangular piece of metal inserted in a slot or groove between a shaft and a hub to prevent slippage.

KEYSEAT—A slot or groove into which the key fits.

KEYWAY—A slot or groove within a cylindrical tube or pipe into which a key fitted into a key seat will slide.

LEAD—The distance a screw thread advances one turn, measured parallel to the axis. On a single-thread screw the lead and the pitch are identical; on a double-thread screw the lead is twice the pitch; on a triple-thread screw the lead is three times the pitch.

LEADER LINES—Two, unbroken lines used to connect numbers, references, or notes to appropriate surfaces or lines.

LEGEND—A description of any special or unusual marks, symbols, or line connections used in the drawing.

LINTEL—A load-bearing structural member supported at its ends. Usually located over a door or window.

LOGIC DIAGRAM—A type of schematic diagram using special symbols to show components that perform a logic or information processing function.

MAJOR DIAMETER—The largest diameter of an internal or external thread.

MANIFOLD—A fitting that has several inlets or outlets to carry liquids or gases.

MECHANICAL DRAWING—*See* DRAWINGS. Applies to scale drawings of mechanical objects.

MIL-STD (military standards)—A formalized set of standards for supplies, equipment, and design work purchased by the United States Armed Forces.

NOTES—Descriptive writing on a drawing to give verbal instructions or additional information.

OBLIQUE DRAWING—A type of pictorial drawing in which one view is an orthographic projection and the views of the sides have receding lines at an angle.

OBLIQUE PROJECTION—A view produced when the projectors are at an angle to the plane the object illustrated. Vertical lines in the view may not have the same scale as horizontal lines.

OFFSET SECTION—A section view of two or more planes in an object to show features that do not lie in the same plane.

ONBOARD PLANS—*See* **SHIP'S PLANS.**

ORTHOGRAPHIC PROJECTION—A view produced when projectors are perpendicular to the plane of the object. It gives the effect of looking straight at one side.

PARTIAL SECTION—A sectional view consisting of less than a half section. Used to show the internal structure of a small portion of an object. Also known as a broken section.

PERPENDICULAR—Vertical lines extending through the outlines of the hull ends and the designer's waterline.

PERSPECTIVE—The visual impression that, as parallel lines project to a greater distance, the lines move closer together.

PHANTOM VIEW—A view showing the alternate position of a movable object, using a broken line convention.

PHASE—An impulse of alternating current. The number of phases depends on the generator windings. Most large generators produce a three-phase current that must be carried on at least three wires.

PICTORIAL DRAWING—A drawing that gives the real appearance of the object, showing general location, function, and appearance of parts and assemblies.

PICTORIAL WIRING DIAGRAM—A diagram showing actual pictorial sketches of the various parts of a piece of equipment and the electrical connections between the parts.

PIER—A vertical support for a building or structure, usually designed to hold substantial loads.

PITCH—The distance from a point on a screw thread to a corresponding point on the next thread, measured parallel to the axis.

PLAN VIEW—A view of an object or area as it would appear from directly above.

PLAT—A map or plan view of a lot showing principal features, boundaries, and location of structures.

POLARITY—The direction of magnetism or direction of flow of current.

PROJECTION—A technique for showing one or more sides of an object to give the impression of a drawing of a solid object.

PROJECTOR—The theoretical extended line of sight used to create a perspective or view of an object.

RAFTER—A sloping or horizontal beam used to support a roof.

RADIUS—A straight line from the center of a circle or sphere to its circumference or surface.

REFERENCE DESIGNATION—A combination of letters and numbers to identify parts on electrical and electronic drawings. The letters designate the type of part, and the numbers designate the specific part. Example: reference designator R-12 indicates the 12th resistor in a circuit.

REFERENCE NUMBERS—Numbers used on a drawing to refer the reader to another drawing for more detail or other information.

REFERENCE PLANE—The normal plane that all information is referenced

REMOVED SECTION—A drawing of an object's internal cross section located near the basic drawing of the object.

REVISION BLOCK—This block is located in the upper right corner of a print. It provides a space to record any changes made to the original print.

REVOLVED SECTION—A drawing of an object's internal cross section superimposed on the basic drawing of the object.

ROOT—The surface of the thread corresponding to the minor diameter of an external thread and the major diameter of an internal thread.

ROTATION—A view in which the object is apparently rotated or turned to reveal a different plane or aspect, all shown within the view.

ROUND—The rounded outside corner of a metal object.

SCALE—The relation between the measurement used on a drawing and the measurement of the object it represents. A measuring device, such as a ruler, having special graduations.

SCHEMATIC DIAGRAM—A diagram using graphic symbols to show how a circuit functions electrically.

SECTION—A view showing internal features as if the viewed object has been cut or sectioned

SECTION LINES—Thin, diagonal lines used to indicate the surface of an imaginary cut in an object.

SHEER PLAN—The profile of a ship's hull, composed of superimposed buttock lines.

SHEET STEEL—Flat steel weighing less than 5 pounds per square foot.

SHIP'S PLANS—A set of drawings of all significant construction features and equipment of a ship, as needed to operate and maintain the ship. Also called ONBOARD PLANS.

SHRINK RULE—A special rule for use by patternmakers. It has an expanded scale, rather than a true scale, to allow for shrinkage of castings.

SILL—A horizontal structural member supported by its ends.

SINGLE-LINE DIAGRAM—A diagram using single lines and graphic symbols to simplify a complex circuit or system.

SOLE PLATE—A horizontal structural member used as a base for studs or columns.

SPECIFICATION—A detailed description or identification relating to quality, strength, or similar performance requirement

STATION NUMBERS—Designations of reference lines used to indicate linear positions along a component such as an air frame or ship's hull.

STEEL PLATE—Flat steel weighing more than 5 pounds per square foot.

STRETCH-OUT LINE—The base or reference line used in making a development.

STUD—A light vertical structure member, usually of wood or light structural steel, used as part of a wall and for supporting moderate loads.

SYMBOL—Stylized graphical representation of commonly used component parts shown in a drawing.

TEMPER—To harden steel by heating and sudden cooling by immersion in oil, water, or other coolant.

TEMPLATE—A piece of thin material used as a true-scale guide or as a model for reproducing various shapes.

TITLE BLOCK—A blocked area in the lower right corner of the print. Provides information to identify the drawing, its subject matter, origins, scale, and other data.

TOLERANCE—The amount that a manufactured part may vary from its specified size.

TOP PLATE—A horizontal member at the top of an outer building wall; used to support a rafter.

TRACING PAPER—High-grade, white, transparent paper that takes pencil well; used when reproductions are to be made of drawings. Also known as tracing vellum.

TRIANGULATION—A technique for making developments of complex sheet metal forms using geometrical constructions to translate dimensions from the drawing to the pattern.

TRUSS—A complex structural member built of upper and lower members connected by web members.

UTILITY PLAN—A floor plan of a structure showing locations of heating, electrical, plumbing and other service system components.

VIEW—A drawing of a side or plane of an object as seen from one point.

WATERLINE—The outline of a horizontal longitudinal section of a ship's hull.

WIRING (CONNECTION) DIAGRAM—A diagram showing the individual connections within a unit and the physical arrangement of the components.

ZONE NUMBERS—Numbers and letters on the border of a drawing to provide reference points to aid in indicating or locating specific points on the drawing.

APPENDIX

GD&T Reference Symbols

GD&T Symbol	Control Type	Name	Summary Description
—	Form	Straightness	Controls the straightness of a feature in relation to its own perfect form
▱	Form	Flatness	Controls the flatness of a surface in relation to its own perfect form
○	Form	Circularity	Controls the form of a revolved surface in relation to its own perfect form by independent cross sections
⌭	Form	Cylindricity	Like circularity, but applies simultaneously to entire surface
⌒	Profile	Profile of a Surface	Controls size and form of a feature. In addition it controls the location and orientation when a datum reference frame is used.
⌒	Profile	Profile of a Line	Similar to profile of a surface, applies to cross sections of a feature
⊥	Orientation	Perpendicularity	Controls the orientation of a feature which is nominally perpendicular to the primary datum of its datum reference frame
∠	Orientation	Angularity	Controls orientation of a feature at a specific angle in relation to the primary datum of its datum reference frame
//	Orientation	Parallelism	Controls orientation of a feature which is nominally parallel to the primary datum of its datum reference frame
⌖	Location	Position	Controls the location and orientation of a feature in relation to its datum reference frame
◎	Location	Concentricity	Controls concentricity of a surface of revolution to a central datum
═	Location	Symmetry	Controls the symmetry of two surfaces about a central datum
↗	Runout	Circular runout	Controls circularity and coaxiality of each circular segment of a surface independently about a coaxial datum
↗↗	Runout	Total runout	Controls circularity, straightness, coaxiiality, and taper of a cylindrical surface about a coaxial datum

Symbol	Meaning
Ⓛ	LMC – Least Material Condition
Ⓜ	MMC – Maximum Material Condition
Ⓣ	Tangent Plane
Ⓟ	Projected Tolerance Zone
Ⓕ	Free State
⌀	Diameter
R	Radius
SR	Spherical Radius
S⌀	Spherical Diameter
CR	Controlled Radius
ⓈⓉ	Statistical Tolerance
[77]	Basic Dimension
(77)	Reference Dimension

Symbol	Meaning
←⊕	Dimension Origin
⊔	Counterbore
∨	Countersink
▽	Depth
⌀ (with arrow)	All Around
↔	Between
✕	Target Point
▷	Conical Taper
◁	Slope
□	Square

The Apprentices Guide to Blueprint Reading

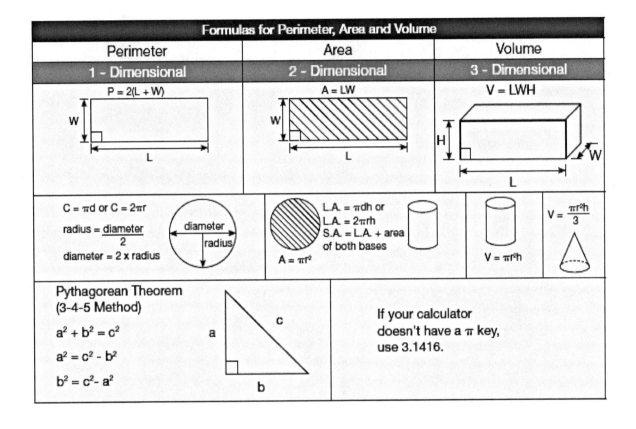

Insert author bio text here Insert author bio text here Insert author bio text here Insert author bio text here I

CPSIA information can be obtained
at www.ICGtesting.com
Printed in the USA
LVHW020427220623
750385LV00007B/282

9 781717 834188